Using MATLAB®
to Analyze and Design
Control Systems

Second Edition

Using MATLAB®
to Analyze and Design
Control Systems

Second Edition

Naomi Ehrich Leonard
Princeton University

William S. Levine
University of Maryland

The Benjamin/Cummings Publishing Company, Inc.

Redwood City, California ■ Menlo Park, California ■ Reading, Massachusetts
New York ■ Don Mills, Ontario ■ Wokingham, U.K ■ Amsterdam ■ Bonn
Paris ■ Milan ■ Madrid ■ Sydney ■ Singapore ■ Tokyo ■ Seoul ■ Taipei
Mexico City ■ San Juan, Puerto Rico

Acquisitions Editor: Tim Cox
Executive Editor: Dan Joraanstad
Assistant Editor: Laura Cheu
Associate Production Editor: Lisa Weber
Copyeditor: Mary Prescott
Cover Designer: Yvo Reizebos
Marketing Manager: Mary Tudor
Manufacturing Supervisor: Casimira Kostecki
Printer: Victor Graphics

The software programs available with this book have been included for their instructional value. They have been tested with care but are not guaranteed for any particular purpose. The publisher and author do not offer any warranties or restrictions, nor do they accept any liabilities with respect to the programs.

Many of the designations used by manufacturers and sellers to distinguish their products are claimed as trademarks. Where those designations appear in this book, and Benjamin/Cummings was aware of a trademark claim, the designations have been printed in initial caps.

Macintosh is a registered trademark of Apple Computer, Inc.
Microsoft is a registered trademark of Microsoft Corporation.
Windows is a trademark of Microsoft Corporation.
MATLAB, SIMULINK, and Handle Graphics are registered trademarks of The MathWorks, Inc.

Library of Congress Cataloging-in-Publication Data
Leonard, Naomi Ehrich.
     Using MATLAB to analyze and design control systems/by Naomi
   Ehrich Leonard and William S. Levine. —2nd ed.
        p.   cm.
     Includes index.
     ISBN 0-8053-2193-4
     1. Automatic control.   2. System design.   3. MATLAB.   I. Levine,
   W. S.   II. Title.
TJ213.L368   1995
629.8'0285'5369—dc20   95-1721
                                                                 CIP

1 2 3 4 5 6 7 8 9 10-VG-99 98 97 96 95

The Benjamin/Cummings Publishing Company, Inc.
390 Bridge Parkway
Redwood City, California 94065

Preface

We wrote this book because we believe that MATLAB can be a very useful aid for learning about and designing control systems. However, we also believe that many people, especially students, would find it difficult to learn to use MATLAB from the MATLAB manuals alone.

Ask yourself how many Nyquist, Bode, and root locus plots you are likely to do by hand. The number is small, yet each such plot provides a great deal of information and insight about a linear system and its possible feedback controls. Software that helps you create such plots in a few minutes makes these tools of control system design more valuable. Using MATLAB, both student and professional can ask and answer questions that previously were ignored because answering them was too time-consuming.

One of the pleasures of writing the book has been the interaction with our friends and colleagues, many of whom helped and encouraged us a great deal. Professor Mark Shayman, Dr. Jiqin Pan, Messrs. Bruce Douglas, Reza Ghanadan, Chujen Lin, John Reilly, Tony Teolis, and Ms. Lei Zhang all read the manuscript, corrected errors, and suggested improvements. Professor Steve Tretter provided very useful advice on Chapter 9. We thank them all.

Our greatest debt is to Mrs. Patricia T. Keehn, who typed the manuscript, formatted the text, and placed the figures. This is a difficult and demanding job that she performed wisely and well.

Finally, we would like to thank Tim Leonard, who consistently provides inspiration and support to the first author, and Shirley Johannesen Levine, who has provided encouragement and moral support to the second author for almost thirty years.

Preface to the Second Edition

One of the best reasons for writing a second edition is that it gives you the opportunity to improve upon your earlier work. One major improvement that we have made in this edition is to include SIMULINK, a tool that is useful for understanding nonlinearity in control systems. Knowing what to expect from a system that is not linear enhances one's understanding and appreciation of linearity and the limitations of linear control design. It is in this context that we have integrated SIMULINK into the book. For instance, we have provided examples of integral windup and conditional stability to demonstrate the use of SIMULINK and to illustrate the effects of actuator saturation.

Another reason for this second edition was to illustrate new features of MATLAB 4.2. In particular, we have provided an introduction to Handle Graphics, a new tool for customizing MATLAB's graphics. We have also changed all of our MATLAB code to use and exploit MATLAB 4.2. Where appropriate, we have described alternatives that will still work with earlier editions of MATLAB.

We have not changed our basic approach which is to present MATLAB by example and as much as possible show how plots can be customized to best display important system and control information. We have updated our comments on how to overcome unavoidable numerical problems that are inherent in control system calculations. In response to comments from users of our first edition we have changed the index.

We want to thank several individuals for their help in preparing this second edition. First and foremost, we thank Mrs. Patricia T. Keehn who again did all of the typing and formatting. There would be no book without her.

We thank Mrs. Margaret Kromer who provided invaluable assistance with TeX and Professor Adam Shwartz who showed us how to manipulate the placement of figures in TeX. We also appreciate the efforts of Laura Cheu and Lisa Weber of Benjamin/Cummings who guided the editorial and production aspects of the book.

Our respective spouses, Tim Leonard and Shirley Johannesen Levine, continue to provide inspiration and support for which we are very grateful.

Contents

Introduction

Just ten years ago, if you wanted to see a plot of the response of a simple linear system to a sinusoidal input, you had only four choices. You could calculate the response and plot it by hand. Alternatively, you could write a computer program to calculate and plot the response. Unless you were a gifted and experienced programmer, the first alternative was much more efficient. A third choice, although few would think to do it, was to actually build the system and measure the response on an oscilloscope. A fourth option was to find a book in which the response was already plotted.

Today, using MATLAB, you can calculate the system response and produce the plot in less than ten minutes. It is equally easy to compute and display the system response for many nonlinear systems using MATLAB and SIMULINK. Moreover, the probability of error is much lower than it would be for a plot made by hand. We believe this power can be used to help learn about systems and control. In fact, we believe that exploitation on a large scale of this improved knowledge, in conjunction with newly available control-system hardware including sensors, actuators, and computers, could even revolutionize the practice of control engineering.

The purpose of this book is to help foment this revolution by teaching you how to use MATLAB and SIMULINK and how to learn from them. Using MATLAB and SIMULINK to replace pencil-and-paper calculation is like using a radial arm saw instead of a hand saw. It is actually easier to learn and use but it is considerably more dangerous. Of course, you are not likely to lose an arm because of a mistake in using MATLAB. But it is possible to make very embarrassing errors. This book provides warnings of the dangers and suggestions for avoiding them.

Our strategy for teaching you how to use MATLAB is based on the belief that MATLAB is basically a language. Trying to learn to use computer software from a manual is usually like trying to learn a language from a dictionary. It is much more effective to learn useful groups of phrases first. The dictionary (the MATLAB manual) becomes more and more useful as your vocabulary improves. Thus, we have filled this book with the computer equivalent of phrases and dialogues.

As with any language, learning phrases is best done by making conversation. For MATLAB this means trying out computer phrases directly on the computer. MATLAB operates interactively and will respond to all your attempts at conversation. Therefore, the more you use MATLAB, the more you will learn. In particular, it is a good idea to try the examples in this book on the computer as you read. Because in MATLAB, as in any language, there is often more than one way to express the same idea, you will be able to come up with alternative ways to solve the problems we have solved in this book. When appropriate, however, guidance is provided regarding preferred MATLAB methods. Throughout this book, inputs to the computer (phrases you might use) and responses from the computer are printed in **bold** type. Bold type is also used for MATLAB commands wherever they appear.

Much of what is called classical control, and is the basis of the first undergraduate course in control as well as this book, deals with linear, time-invariant, single-input, single-output (SISO) systems and is overwhelmingly graphical. MATLAB allows you to generate Bode, Nyquist, and Nichols plots, root locus plots, pole-zero plots, and plots of transient responses with almost no effort. Thus, you can focus on the information in all of these graphical displays and not on the tedium of creating them by hand. In fact, there are many exercises and examples in the book that are intended to teach aspects of control that could not be taught when students needed roughly an hour to produce one plot.

Similarly, nonlinearities have largely been ignored in the first undergraduate course in control. SIMULINK, by enabling students to simulate nonlinear systems rapidly and efficiently, makes it possible to include important nonlinear effects in the first course in control. For example, see the discussion of conditional stability in Chapters 4, 5, and 6.

The exercises are also meant to address another issue. Passive learning is not very effective. It is important to be engaged by the material being studied. It is important to think about the information being displayed and not just watch it go by. The exercises require thought and are essential for learning the material.

While this book focuses on using MATLAB to study SISO systems, many of MATLAB's commands support the study and design of multiple-input, multiple-output (MIMO) systems. The interested reader will find, by referring to the MATLAB manual or the on-line help facility, that using MATLAB for MIMO systems requires only minor changes and additions to the examples provided in this book. Additionally, MATLAB can be used to investigate more advanced modern control techniques such as the linear quadratic regulator (LQR) optimal controller. This book provides a basis for further exploration of these advanced topics.

The fact that this book is intended to teach you to use MATLAB and to learn from MATLAB imposes some structure on the book. Chapters 1 and 2 are devoted to showing you the fundamentals and basic plotting capabilities of MATLAB. This includes a brief introduction to Handle Graphics in Chapter 2. The remainder of the book is much more focused on control. Because control theory and design are such graphical subjects, this part of the book emphasizes using MATLAB to learn quickly what used to take years to learn. However, this book is a supplement rather than a replacement for a conventional textbook on controls. Our book follows the outline of *Control Systems Engineering*, second edition, by Norman S. Nise (Benjamin/Cummings Publishing Co., 1995), but it could be used with almost any controls textbook.

This book is based on MATLAB Version 4.2a with the MATLAB Control System Toolbox Version 3.0b and Version 1.3a of SIMULINK for Unix-based systems, MS-DOS compatible personal computers, and Macintosh computers. Every attempt has been made to remain downward compatible with MATLAB Versions 3.5, 4.0, and 4.1. Commands that will function properly in Version 3.5 are given where MATLAB 4.2 is not downward compatible.

In general, the MATLAB commands used are machine independent. However, commands to perform tasks such as file editing and printing will differ from computer to computer. This issue is discussed further in Chapters 1 and 2.

1 MATLAB FUNDAMENTALS

This chapter introduces the basic MATLAB commands and their use. The approach in this chapter, as it will be throughout the book, is to illustrate the use of the MATLAB commands by example. Because this chapter deals with the MATLAB fundamentals, additional examples of the use of many of these commands will be found in subsequent chapters. Although many examples are provided in this chapter, no attempt is made to cover every possible use or variant of the commands. For additional information on the individual MATLAB commands, we refer you to the MATLAB User's Guide and MATLAB Control System Toolbox guide or the on-line help facility described in Section 1.2.

1.1 Getting Started

MATLAB is started by entering **matlab** at the system prompt or by clicking on the MATLAB icon, depending on the type of machine you are using. Once invoked, MATLAB will clear the screen, provide some introductory remarks, and produce the MATLAB prompt >>.

For the most part, MATLAB commands are independent of the type of machine and operating system you are using. However, the way that MATLAB interfaces with your computing environment varies dramatically from machine to machine. As a result, use of features such as printing and command-line editing (described in the following) depends on your machine. Additionally, as you will learn in Section 1.3, calling files created by a text editor external to the MATLAB environment is an essential part of effective MATLAB use. As there is a multitude of editors for the various machines, your editing commands will also be machine specific and editor specific. Throughout the book we will identify important MATLAB features that are machine dependent. For these features check with your local experts for further information.

To exit from MATLAB, type **quit** or **exit** at the MATLAB prompt, followed by either the "enter," "return," or "carriage return" key. On some machines with a windows environment such as the Macintosh you can alternatively pull down the **file menu** and click on **quit** to exit from MATLAB.

From here on we will not repeat the instruction to depress the "enter," "return," or "carriage return" key after a MATLAB command is typed. It is understood that you must do this to execute a command; i.e., MATLAB does not respond to a line of commands until one of those keys is depressed.

1.2 Fundamental Expressions

Working in the MATLAB environment is straightforward because most commands are entered as you would write them mathematically. For example, entering the following simple expression

```
>> a = 4/3
```

yields the MATLAB response

```
a =
   1.3333
```

and assigns to variable **a** the value of 4 divided by 3. In general, it is a good idea to use appropriate and memorable names for variables. MATLAB recognizes the first 19 characters of a variable name and requires only that the first character in a variable name be a letter. MATLAB is case sensitive, so, for example, **a** and **A** are two different variables. All MATLAB commands are written in lowercase letters.

If you do not care to create a new variable but want to know the value of an expression, you can type the expression by itself, e.g.,

```
>> 4/3
```

which yields

ans =
     1.3333

where **ans** is a MATLAB-generated dummy variable that stands for "answer." Alternatively, if you prefer to create a new variable but do not wish to see the MATLAB response, type a semicolon at the end of the expression. For example,

>> **b** = 4+7;

will create a new variable **b** whose value is 11, but MATLAB will not display the value of **b**. The semicolon is very useful when large vectors or matrices are defined or computed in intermediate steps of a computation. Watching MATLAB display a 100×100 matrix that you do not care about can be fairly tedious. Moreover, you can check the value of a variable at any time by entering the variable name at the prompt as follows:

>> **b**
b =
     11

This is an extremely useful feature, especially when you are debugging a sequence of expressions.

Expressions, as opposed to variables, can be made up of sequences of numbers, operators, functions, and variables that have been previously defined. For example, since **a** and **b** have already been defined, we can do the following:

>> **c** = a*(b-1)
c =
     13.3333

Note that blank spaces can be used to improve readability. If you are typing in an expression that does not fit on one line, use ellipsis dots (three or more periods) at the end of the line and continue typing on the next line, e.g.,

>> **p** = 1+2+...
     3+4+6;

Arithmetic operators are the same as those commonly used except that * represents multiplication, \ performs left division, and ^ is the power operator. For example, typing

>> **p^ 2**

causes MATLAB to return

ans =
        256

Left division is defined like right division except that the order of dividend and divisor is swapped, i.e., $a \backslash b = (a)^{-1}b$. For a and b scalars, $a \backslash b = b/a$.

MATLAB performs operations in the following order:

^	power operator
*	multiplication
/ and \	division
+	addition
−	subtraction

Precedence of like operators proceeds from left to right, but parentheses can be used to affect the order of operation. The following three examples illustrate these precedence rules:

>> **1+2^ 3/4*2**
ans =
    5
>> **1+2^ 3/(4*2)**
ans =
    2

>> **(1+2)^ 3/(4*2)**
ans =
    3.3750

MATLAB has several predefined variables. These include **i** and **j**, both of which denote $\sqrt{-1}$. Using **i** or **j** to generate complex numbers can be very convenient. However, predefined variables can be overwritten, so be careful using the variable names **i** and **j** to mean something other than $\sqrt{-1}$. For example, many people use i and j as indices for vectors and matrices. As a result, they often use **i** or **j** this way in MATLAB, e.g., by entering **i=1** to reassign **i** to be equal to 1 instead of $\sqrt{-1}$. We recommend that you avoid changing the values of predefined variables. However, if you choose to do so, it is good practice to reset them after use with the **clear** command. For example, to retrieve the predefined value of $\sqrt{-1}$ for **i** type **clear i**.

Other predefined variables are

pi, which stands for π
Inf, which stands for ∞
NaN, which stands for not a number (e.g. 0/0)

MATLAB will return ∞ or **NaN** when you divide by zero. When this happens execution of your calculation will not be terminated; however, MATLAB will warn you of the problem:

```
>> d = 4/0
Warning:  Divide by zero
d =
     ∞
```

In Version 3.5, ∞ would be represented as **Inf**. Similarly, MATLAB returns **NaN** when you attempt to perform certain undefined calculations, e.g.,

```
>> dd = Inf/Inf
dd =
     NaN
```

Note that in this case no warning is given. In general, you should try to avoid generating **Inf** and **NaN** because they can lead to meaningless results in your calculations. When they are unavoidable you should deal with them explicitly, for example, as illustrated in Chapter 2, Section 2.3.

MATLAB also has a large variety of functions that are easily incorporated into expressions. The simplest functions have one input argument and one output argument, such as the **sqrt** function, which returns the square root of the input argument. The input argument goes inside a pair of parentheses. For example,

```
>> y = sqrt(1+4*i)
```

returns

y =
      1.6005 + 1.2496i

Notice the way in which the complex number $1 + 4i$ was entered.

1.3 On-line Help, Format, and Save

As we said in the introduction, we don't think you learn a language like MATLAB from a list of its variables, operators, and functions. So rather than listing the remaining simple MATLAB functions, we discuss how you can explore them further using the **help** command. Additionally, we briefly describe some useful MATLAB features (**format**, **save**, **load**) and then proceed to more sophisticated uses of MATLAB.

A dictionary becomes very helpful once you know some of the basics of a language. Entering **help** by itself at the MATLAB prompt displays a list of MATLAB help topics (MATLAB functions in Version 3.5). These include, for example, "ops — Operators and special characters," "elfun — Elementary math functions," and "plotxy — Two-dimensional graphics." To get more information and lists of MATLAB functions for a particular topic, type **help** followed by the name of the topic. For instance, typing **help elfun** will give a list of elementary math functions categorized as trigonometric, exponential, complex, and numeric. This list will give a brief description of each function. To get more information about an individual command, type **help** followed by the command name.

These lists are easy to use even if you want to perform an operation using MATLAB but do not know which command to use. Most of the MATLAB commands have fairly mnemonic names, so skimming the topics and lists until you find some likely candidates usually works. You can also use the command **lookfor** to search through the lists using a keyword.

Suppose you want to compute e^{-1+3i}. Scanning the list of MATLAB topics for the appropriate exponential command leads you to "elfun" for elementary functions. Scanning the list of elementary functions clearly leads you to the command **exp**. To find out more about **exp**, type

>> help exp

MATLAB returns a detailed description of **exp**, which indicates that **exp** is the command you needed.

When using the help facility, you will find that MATLAB may return more information than can fit on your command window. To control MATLAB output so that you can view it page by page, use the **more** command. Typing **more on** enables output control. In this case, MATLAB will display only one page of output. To see the next line of output hit the return key. To advance to the next page of output hit the spacebar. Hitting the "q" key will stop the outputting and return you to the MATLAB prompt. Type **more off** to disable paged output. This controlled output feature is also very useful when you want to view large vectors or matrices page by page.

When you enter **help** by itself, MATLAB will also display a list of all of your directories defined in the MATLAB search path environment variable MAT-LABPATH. These are directories where you will have created and stored your own MATLAB script files and functions (see Sections 1.4 and 1.7 for discussions of how to create such files and functions). If you have created a table of contents file called *Contents.m* in each of your directories, MAT-LAB will display next to the corresponding directory name the first line of *Contents.m* that starts with **%**. Typing **help** followed by the name of the directory will give you a listing of all the lines of *Contents.m* that begin with **%**. It is there that you may want to list the names of the files in your directories with short descriptions. This allows you to scan the dictionary of your own work quickly. Furthermore, if you use **help** followed by the name of one of your script files or functions, MATLAB will respond with information on your command if you have included this information in your original file. Sections 1.3 and 1.6 describe how to include this information. Note that MATLABPATH is predefined for you. To change it, consult the MATLAB manual or your local experts.

You should also be aware that there are several MATLAB commands that are not described in the MATLAB manual. Type **help README** or edit the file *README.m* to see a list of these commands. For help on the on-line help facility type **help help**.

In all of the preceding examples, MATLAB displays its numerical responses with five decimal digits. This is MATLAB's default display. However, you can change the type of MATLAB numerical display using the **format** command. For example, to see a five-digit floating point display (scientific notation) enter **format short e**. The new display will look as follows:

```
>> c
c =
     1.3333e+01
```

Typing **format** returns the display to the default.

It is important to keep in mind that the number of decimal digits that MATLAB displays has no relation to the number of digits that are accurate for your individual problem. That is, if you know that your results will be accurate to only two significant decimal digits, then you should use the appropriate MATLAB display or round off the MATLAB response to two significant digits. MATLAB is exceptionally good about reporting numerical problems to its user, but no computer program knows the accuracy of the numbers it is given. For example, suppose you know that a rectangular solid has volume 0.01 m^3 with height measured as 0.12 $\pm$ 0.01 m and width measured as 0.06 $\pm$ 0.01 m. We can find the depth of the rectangular solid using MATLAB as follows:

```
>> V = 0.01;
>> height = 0.12;
>> width = 0.06;
>> depth = V/(height*width)

depth =
     1.3889
```

Based on the accuracy of the measurements used in the calculation, the depth should be reported as 1.4 m even though MATLAB provides additional decimal digits.

Before quitting MATLAB you might like to save the variables you generated during the current work session, particularly if you have performed a long sequence of calculations and want to continue working with the final results at another time. The **save** command will save all of the current user-generated variables in a file called *matlab.mat*. Alternatively, **save thewhales** will save the variables in a file *thewhales.mat*. Individual variables can be saved too. For example, **save learnyour a b c** will save variables **a**, **b**, and **c** in the file *learnyour.mat*. The command **load** is used to restore the variables at a subsequent MATLAB session, e.g., **load learnyour**. The **who** and **whos** commands will give you a list of your current variables. The **clear** command erases your variables.

MATLAB also provides some command-line editing facilities that allow you to correct typing errors and recover previously entered commands in the MATLAB environment. Typically, the arrow keys and others, such as the delete key, can be used for editing. However, the details of the command-line editing tool vary from computer to computer, so check with your manual or with someone familiar with your system for more information on this facility.

1.4 Creating Script Files

It is much more convenient to use script files than to enter commands line by line at the MATLAB prompt. A script file is an ASCII file (regular text file) that contains a series of MATLAB commands written just as you would enter them in the MATLAB environment. Statements that begin with a **%** are considered to be comments and are ignored by MATLAB. The script file is created outside of MATLAB with any available text editor or word processor (check with your instructor or other local experts for further information on text editing for your computer system). Each script file should have a name that ends in ".m" (optional on Macintosh computers). The commands in the script file are executed in the MATLAB environment simply by entering the name of the script file without the ".m." For example, suppose the text file *magphase.m* contains the following statements:

```
% magphase.m:  example m-file to
% compute the magnitude and phase of G at w=1.
w=1;
G=1/(j*w + 2);
mag=abs(G)
phase=atan(imag(G)/real(G))
```

Then typing **magphase** at the MATLAB prompt will yield the following MATLAB response:

```
mag =
     0.4472
phase =
    -0.4636
```

which are the magnitude and phase of the transfer function $G(j\omega) = 1/(j\omega + 2)$ evaluated at $\omega = 1$. Entering **help magphase** gives you back the text in the comment lines at the beginning of the file:

magphase.m: example m-file to
compute the magnitude and phase of G at w=1.

It should be obvious that it is very desirable to include at least some brief comments as a header to each m-file you create.

In a windowing environment such as that on a Macintosh computer, on a PC with Microsoft Windows, or on a Unix-based workstation with X-windows, you can view the text editor window simultaneously with the MATLAB command window. This means that you can use the two windows to edit a script file repeatedly and run it in MATLAB without ever quitting MATLAB. However, on a PC, if you are using MATLAB Version 3.5, you will not be able to use Microsoft Windows. As a result, you will have only one available window, so you may have to quit MATLAB to edit your script file. The ! command allows you to run a program outside MATLAB without quitting MATLAB; i.e., ! temporarily returns you to your normal operating system. Therefore, if your text editor uses up a relatively small amount of memory, you can run it without quitting MATLAB. For example, to edit *magphase.m* on a PC with the DOS editor *edit*, type **!edit magphase.m** at the MATLAB prompt. Proceed with your editing, and when you are finished quit your editor to return to MATLAB.

On the other hand, if you are using an editor that requires a large amount of memory on a PC without Microsoft Windows, you will have to quit MATLAB to edit your script file and then restart MATLAB to run your file. The MATLAB batch file *matlab.bat* can be used to streamline this process. First edit *matlab.bat* so that it contains the appropriate command to invoke your editor. Then when you are in MATLAB just type **edit** when you want to work on a script file. MATLAB will respond by saving all your current variables, quitting MATLAB, and calling your editor. When you are done editing, MATLAB will be restored along with all the variables that were saved. Type **help edit** for more help on this feature. The command **edit** is not available in MATLAB Version 4.2.

Because MATLAB treats script files exactly as if they are command sequences, all variables currently in the MATLAB workspace can be used by the script file commands, and similarly all variables created by the script file are available for use after the script file has been run. For example after running *magphase.m* we can examine **G** as follows:

>> **G**
G =
        0.4000-0.2000i

When any script file name is entered at the MATLAB prompt, MATLAB searches in several places to find the corresponding file. Suppose you have entered **sample** in the hope of running a script file you have written and named *sample.m*. MATLAB first checks to see if **sample** is a variable and then if it is a built-in function. Next, MATLAB looks in the current directory for the file *sample.m*. If the file is not there, MATLAB then looks in the directories specified by the environment variable MATLABPATH (check with your local experts if you need to adjust MATLABPATH). Consequently, to avoid confusion, make sure your m-files do not have the same name as previously defined variables, m-files, or MATLAB functions. Also, be aware that MATLAB will not find your file at all if it is neither in the current directory nor in a directory specified in MATLABPATH.

The script file *startup.m* is run automatically when MATLAB is invoked. Therefore, you might want to create or edit the file *startup.m* to include commands that you would like run every time you invoke MATLAB. For example, you might include the command **format short e** to set the numerical display automatically.

1.5 Matrices, Vectors, and Polynomials

Matrices are entered into MATLAB by listing the elements of the matrix and enclosing them within a pair of brackets. Elements of a single row are separated by commas or blanks, and rows are separated by semicolons or carriage returns. For example,

>> **A = [1 2; 3 4]**

yields the MATLAB response

A =
        1   2
        3   4

and

```
>> A = [1,2
        3,4]
```

produces the same result.

Matrix elements can be any MATLAB expression; however, MATLAB recognizes only rectangular matrices; i.e., the matrix must have the same number of columns in each row. To find the dimensions of a matrix use the **size** command, e.g.,

```
>> size(A)
ans =
      2    2
```

Individual matrix elements can be referenced using indices enclosed within parentheses. The first index identifies the row number, and the second index identifies the column number. For instance, to change the second element in the second row of matrix **A** to 5, type

```
>> A(2,2) = 5
A =
      1    2
      3    5
```

If you add an element to a matrix beyond the existing size of the matrix, then MATLAB automatically inserts zeros as needed to maintain a rectangular matrix:

```
>> A(3,3) = 6
A =
      1    2    0
      3    5    0
      0    0    6
```

Since a vector is simply a $1 \times n$ or an $n \times 1$ matrix, where n is any positive integer, you can generate vectors in the same way as matrices:

```
>> v = [sin(pi/3)   -7^3   a+1]
v=
      0.8660      -343.0000   2.3333
```

Alternatively, special vectors can be created using the **:** operator. The command **k = 1:10** generates a row vector with elements from 1 to 10 with increment 1. Any other increment can be applied with a second : as follows:

```
>> knew = 1:0.25:2
knew =
1.0000   1.2500   1.5000   1.7500   2.000
```

The command **logspace(x,y,n)** creates vectors with **n** elements that are spaced in even logarithmic increments between 10^x and 10^y. This command is particulary useful when you want to generate data for plotting on a logarithmically scaled graph such as a Bode plot (described further in Chapter 6). The command **linspace** is similar to **logspace** except that the vector elements are spaced linearly.

Using the simple commands described thus far you can easily manipulate both matrices and vectors. For example, to add a row onto matrix **A** we type

```
>> A = [A; [7 8 9]]
A =
     1     2     0
     3     5     0
     0     0     6
     7     8     9
```

To extract the submatrix of **A** that consists of the first through third columns of the second through third rows, use vectors as indices as follows:

```
>> B = A (2:3,1:3)
B =
     3     5     0
     0     0     6
```

The : command as an index by itself indicates that all elements should be selected; i.e., **A(1,:)** will return the entire first row of matrix **A**. Typing **A(:)** will return all the elements of **A** strung out into a column vector.

Matrices can also be manipulated using vectors that contain elements equal to 0 and 1 only. These 0-1 vectors are most frequently generated using relational operators, such as > or <, described in the next section. If you use a 0-1 vector as a matrix index, rows or columns corresponding to the 1's are selected while those corresponding to the 0's are not selected, such as,

```
>> B = B (:,[0 1 1])
B =
      5    0
      0    6
```

MATLAB has commands for generating special matrices. For example, you can create a diagonal matrix with the **diag** command using a vector containing the diagonal elements as the input argument, such as,

```
>> D = diag([1 2])
D =
      1    0
      0    2
```

Use **diag** with a second input argument to put the vector on a diagonal other than the main diagonal. Another useful special matrix is the identity matrix. **eye(4)** creates a 4×4 identity matrix. **eye(B)** generates an identity matrix of the same dimension as matrix **B**, i.e. 2×2, in MATLAB Version 3.5. This has been replaced by **eye(size(B))** in MATLAB Version 4.1 and higher, although, at this writing, both will still work. The **zeros, ones,** and **rand** commands work similarly to **eye** and make matrices with elements equal to zero, elements equal to one, and random elements (uniformly distributed on the interval 0 to 1), respectively. These commands can also be used to create nonsquare matrices. For example, **zeros(2,4)** generates a 2×4 matrix of zeros.

As an additional example, to generate the state-space canonical form for a matrix A given that the characteristic polynomial of A is $s^4 + 5s^3 + 3s^2 + 2s + 1$ we type

```
>> Acanonical = [zeros(3,1) eye(3); -1 -2 -3 -5]
Acanonical =
      0    1    0    0
      0    0    1    0
      0    0    0    1
     -1   -2   -3   -5
```

Alternatively, MATLAB provides the command **canon** for state-space transformations to canonical form. We discuss this further in Chapter 7.

Polynomials are described in MATLAB by row vectors with elements that are equal to the polynomial coefficients in order of decreasing powers. For example, to enter the polynomial $p = s^2 + 5s + 6$ type **p** = [**1 5 6**]. Zero coefficients must be included to avoid confusion; i.e., $q = s^3 + 5s + 6$ is entered as **q** = [**1 0 5 6**]. A polynomial can be evaluated using the **polyval** command. For example,

```
>> polyval(p,1)
ans =
     12
```

gives the value of the polynomial **p** at $s = 1$. The **roots** command is a convenient way to find the roots of a polynomial, e.g.,

```
>> r = roots(p)
r =
      -3
      -2
```

Similarly, you can construct polynomials from roots,

```
>> t = poly([-4 -5])
t =
    1   9   20
```

If you use a matrix as the input argument to **poly**, then the characteristic polynomial equation is returned:

```
>> poly(B)
ans =
      1   -11   30
```

The command **conv** multiplies two polynomials and **deconv** divides two polynomials. The script file *polyroly.m* shown below illustrates these commands in use.

polyroly.m
```
% polyroly
% Example of multiplying and dividing polynomials.
f1=[1 3 2];              % f1=s^2 +3s +2
f2=[1 3];                % f2=s+3
f3=[1 2];                % f3=s+2
```

```
g=conv(f1,f2);                    % g=(s^2+3s+2)(s+3)
h=deconv(g,f3)                    % h=(s^2+3s+2)(s+3)/(s+2)
```

Note the use of **%** to begin a comment. Everything following **%** on a line is ignored by MATLAB. Typing **polyroly** at the MATLAB prompt yields the following response:

```
h =
     1    4    3
```

In this example **f3** divides **g** with no remainder. In general, when used with one output argument, the command **deconv** will return only the quotient even if the remainder is nonzero. However, if two output arguments are specified as shown below, MATLAB will return both the quotient **q** and the remainder **r**:

```
>>[q,r] = deconv(f1,f2)
q =
     1    0
r =
     0    0    2
```

1.6 Matrix Operations and Functions

MATLAB performs matrix arithmetic as easily as it does scalar arithmetic. To add two matrices, simply type

```
>> B+D
ans =
     6    0
     0    8
```

Similarly, to multiply two matrices do as you would for scalars:

```
>> B*D
ans =
     5    0
     0    12
```

Dividing by matrices is also straightforward once you understand how

MATLAB interprets the divide symbols **/** and ****. Suppose you want to solve for x in the equation $Px = Q$. To express the solution $x = P^{-1}Q$ in MATLAB use left division as **x = P\Q**. Now suppose you want to solve $yP = Q$ for y. The solution to this problem is $y = QP^{-1}$, which you can write in MATLAB using right division as **y = Q/P**.

Although MATLAB needs no special instructions to multiply or divide matrices, it does require that the inner dimensions of the two matrices being multiplied or divided be the same. The exception is multiplying or dividing a matrix by a scalar, which is a valid calculation. MATLAB will tell you when you try to multiply matrices having incompatible dimensions. For instance, running *mistake.m*:

mistake.m
```
% mistake:  Shows what happens when you try to multiply
% matrices having incompatible dimensions.
X=eye(2);
Y=[1 2; 3 4; 5 6];
Z=X*Y;
```

yields the MATLAB error response,

```
???  Error using ==> *
Inner matrix dimensions must agree.
Error in ==> mistake.m
On line 5 ==> Z = X*Y;
```

To debug *mistake.m* enter **size(X)** and then **size(Y)** at the MATLAB prompt to check the dimensions of the matrices **X** and **Y**. Of course, you will find that **X** is a 2×2 matrix and **Y** is a 3×2 matrix. Therefore, the last statement in *mistake.m* is in error because the inner dimensions of **X** and **Y** are not the same. Multiplication can be performed, however, if we use the transpose of **Y** because the tranpose of **Y** is a 2×3 matrix. The **'** command computes the complex conjugate transpose (the adjoint) of a matrix. For a real matrix this is simply the transpose. Thus, replacing the last line of *mistake.m* with **Z = X*Y'**; and rerunning *mistake.m* will give a valid result. You should probably rename the m-file as well!

The power operator **^** also operates on the matrix as a whole as long as the matrix is square. For example, **X^ 2** is equivalent to **X*X**. However, **Y^ 2** is illegal because **Y** is not a square matrix.

MATLAB includes many functions that perform some operation on a matrix, such as **det(X)** and **inv(X)**, which produce the determinant and inverse of **X**, respectively. **rank(X)** determines the rank of matrix **X**, **eig(X)** returns the eigenvalues of **X** in a column vector, and **expm(X)** computes e^X. Note that many of these functions require that the input argument be a square matrix. To explore other functions and to learn more about using these functions use **help** as described in Section 1.3.

It is important to understand that some matrices are poorly conditioned, meaning that many calculations involving the matrix can produce results containing substantial numerical errors. The **hilb** command generates a square Hilbert matrix, a good example of an ill-conditioned matrix:

```
>> H=hilb(3)
H =
      1.0000      0.5000      0.3333
      0.5000      0.3333      0.2500
      0.3333      0.2500      0.2000
```

The basic problem is that numerical solutions to problems involving ill-conditioned matrices are highly sensitive to perturbations in the input data. For example, suppose we want to solve $y = Hx$ for x when y=[1.83 1.08 0.78]′. We enter the following:

```
>> x = H\y
x =
      0.9900
      1.0800
      0.9000
```

which looks fine. However, suppose we repeat the problem with a small perturbation **e** added to the data of **y**:

```
>> e = [0.01 -0.01 0.01]';
>> x = H\(y+e)
x =
      1.7400
     -3.0000
      4.8000
```

As you can see, the small perturbation resulted in very dramatic changes in the solution **x**.

It is a good idea to use the **cond** command to check whether matrices you use in your calculations are ill-conditioned. **cond(A)** provides the condition number of matrix **A**. The condition number is a quantitative measure of how ill-conditioned a matrix is. For example, **H** has a high condition number:

```
>> cond(H)
ans =
      524.0568
```

The logarithm (base 10) of the condition number gives you a rough idea of how many decimal places the computer might lose to round-off errors. Therefore, calculations using **H** can be expected to lose two or three decimal places of accuracy. Because poorly conditioned matrices are not uncommon, it is also good practice to remember a sample ill-conditioned matrix and use it to test any matrix manipulation routines you write. An additional example of an ill-conditioned matrix is

```
>> ill_matrix = [1 10000; 0 1]
ill_matrix =
            1       10000
            0       1
>> cond(ill_matrix)
ans =
      1.0000e+08
```

At times you may want to consider a matrix as simply an array of numbers and operate on the array element by element. Specifically, you will create arrays to represent tables of data that you want to manipulate. For instance, tables of data generated by experimentation or other software can be loaded into MATLAB for manipulation using the command **load** (with the suffix **-ascii**). MATLAB provides different ways of using functions to operate on arrays instead of matrices. For example, suppose you have a table of data that you have entered as an array called **Data**. Now suppose you would like to perform a root-mean-square calculation and need to find the square of each element in **Data**. Using a **.** you can convert arithmetic matrix operations into element-by-element operations (addition and subtraction are the same in either case). Specifically, preceding the operator with the **.** indicates array operations. To square each element in the array, type **Data.^2**. Similarly, to multiply two arrays **R** and **S** (of the same dimensions) element by element, type **R.*S** as follows:

```
>> R = [4      5
           0      1];
>> S = [2      3
           4      6];
>> R.*S
ans =
     8      15
     0       6
```

Some MATLAB functions automatically operate element by element on an array. For example, **exp(X)** will return an array with each element equal to the exponential of the corresponding element of **X**. [If you actually want e^X you have to use **expm(X)**]. The trigonometric functions and other elementary mathematical functions such as **abs**, **sqrt**, **real**, and **log** also operate element by element.

MATLAB's relational operators and logical operators also work on an element-by-element basis. Relational operators compare two scalars and produce a 1 if the operation is true and a 0 if it is false. For example, if you enter **t=17>55**, MATLAB will respond with **t=0**. When used with two matrices, relational operators compare corresponding matrix elements. For example, **L = D <= X** will check every element of **D** against the corresponding (by location) element of **X**. If the element of **D** is less than or equal to the corresponding element of **X**, the corresponding element of **L** will be 1. Otherwise, the corresponding element of **L** will be zero, as seen below.

```
>> L = D <= X
L =
     1      1
     1      0
```

The logical operators **&** for logical AND, **|** for logical OR, and **~** for logical NOT all return 1 for logical TRUE and 0 for logical FALSE. MATLAB considers all nonzero values to be logical TRUE. For example,

```
>>R&S
```

yields

ans =

     1   1

     0   1

because all elements of both **R** and **S** are logical TRUE except the lower left element of **R**. Similarly, we get

>>~L

ans =

     0   0

     0   1

To learn more about the relational operators type **help relop** or, e.g., **help <=.**

1.7 Creating Functions

When you have perfected a sequence of commands that performs a useful function, you might want to convert it into a new function command and create a task-independent extension of MATLAB. To do so you need to create a function file. Function files are m-files that are very similar to script files. The first major difference is that the first line of a function file begins with the word **function**, followed by a statement identifying the name of the function and the input and output arguments in the form:

function [*output arguments*]=function name (*input arguments*).

For example, suppose you want to create a new function called **rms** that computes the root mean square of a list of numbers. The first line of your function file might look like **function y = rms(v)**. The next several lines should be comment lines describing how to use **rms**. Later, when you type **help rms**, MATLAB will respond with these comment lines.

The remaining lines of the function file should look as they would in a script file. Remember that the input argument **v** will be defined when the function is called. Also, remember to use the ; command at the end of each line. Unless you are debugging your function, it is inappropriate to display the intermediate steps of a function. The function file to produce the **rms** command might look as follows:

rms.m

```
function y=rms(v)
% rms    Root mean square
% rms(v) returns the root mean square of the
% elements of column vector v. If v is a matrix then
% rms(v) returns a row vector such that each element
% is the root mean square of the elements in the
% corresponding column of v.
vs=v.^2;
s=size(v);
y=sqrt(sum(vs)/s(1));
```

Try creating and using **rms** on an input of your choice. Note that whereas a script file name can only be entered by itself at the MATLAB prompt (or as a line in another script file), a function can be called when needed. For example, if **p** is a vector or matrix that has been specified, you could use **rms** in an expression such as **differ=max(p)–rms(p)**. Also try **help rms** to confirm implementation of the **help** command.

The second major difference between function files and script files is that the variables generated in function files are local to the function, whereas variables in script files are global. This means that the variables **y, v,** and **vs** in our example are not available after the function has been used. Therefore, if you enter **z=rms(p)**, only **p** and **z** are available after running.

However, for debugging or demonstration purposes it may be necessary to examine the local variables. MATLAB makes available several functions that aid in debugging or presenting your function files. For example, the **echo** command displays each command in your file on the screen as it is performed. The **pause** command forces a routine to stop until you press a key. Even more powerful, **keyboard** stops the routine and allows the user to enter any series of commands. The routine is continued when **return** (spelled out) is entered. That is, type the word **return** and then strike the return key. The **input** command allows the function to request information from the user. Note that all of these commands can be used similarly in script files.

The commands **echo, pause, keyboard,** and **input** are all available with MATLAB Version 3.5. Using them for debugging requires editing and re-editing your m-files. However, starting with Version 4.0, MATLAB has a new set of debugging tools that can be used without editing your m-file. These tools are to be used only to debug function files and not for script files. The MATLAB debugger allows you to set break points, check variables in various workspaces, step through functions, etc. For more information type **help debug**.

MATLAB also makes available the control flow commands **for, while**, and **if**. These are statements similar to those used in many computer languages to produce conditional statements or loops. For example, the function **init** defined below uses the **for** command to initialize an $m \times n$ matrix such that each element is the sum of its indices.

init.m
```
function M=init(m,n)
% init:  Creates an m×n matrix
% such that each element is the sum of its indices.
for k=1:m
    for l=1:n
        M(k,l)=k+l;
    end
end
```

Every **for, while**, and **if** statement must be matched with an **end** statement. The **break** command can be used to terminate the execution of a loop prematurely.

The **if** command can be used in conjunction with the **nargin, nargout**, and **error** functions to perform error checking of a function. Inside a function file, **nargin** and **nargout** are equal to the numbers of input and output arguments, respectively, that were used in the function call. The command **error('message')** returns control to the keyboard and displays **message**. Thus, to ensure that our function **init** is run only if two input arguments are given, insert the following lines immediately after the comment lines of *init.m*:

```
if (nargin ~= 2)
    error('Incorrect number of input arguments')
end
```

Try running **init** with the correct and incorrect number of input arguments. Note that the error message is enclosed between single closed quotation marks. Use **help** (topic is **"lang"**) to find out more about **for, while, if, break**, and **error**.

In addition to creating your own functions, you should take advantage of the fact that you can customize many of the existing MATLAB functions. Because the majority of the MATLAB functions are provided in m-files that you can access, you can create a new function by making a copy of any existing m-file and editing it. The MATLAB functions also provide good examples of MATLAB code. Use the command **type** followed by a MATLAB command name to see the contents of the associated m-file. Browsing through these files can give you ideas that you can use to make MATLAB perform for you.

1.8 Exercises

1. Consider

$$A = \begin{bmatrix} -5 & 1 & 0 \\ 0 & -2 & 1 \\ 0 & 0 & 1 \end{bmatrix}, \quad b = \begin{bmatrix} 0 \\ 0 \\ 1 \end{bmatrix}, \quad c = \begin{bmatrix} -1 & 1 & 0 \end{bmatrix}.$$

   (a) Suppose $Ax = b$. Find x.

   (b) Suppose $yA = c$. Find y.

   (c) $G(s) = c(sI - A)^{-1}b$. Find $G(0)$ and $G(1)$.

   (d) Define $C_M = [b \ Ab \ A^2b]$. Find the rank of C_M.

   (e) Now consider an arbitrary $n \times n$ matrix A and $n \times 1$ vector b. Let $C_M = [b \ Ab \ A^2b \ldots A^{n-1}b]$. Write a script file that computes the rank of C_M.

2. Consider the function

$$H(s) = \frac{n(s)}{d(s)}$$

   where
$$n(s) = s^3 + 6.4s^2 + 11.29s + 6.76$$
$$d(s) = s^4 + 14s^3 + 46s^2 + 64s + 40$$

   (a) Find $n(-12), n(-10), n(-8)$. Find $d(-12), d(-10), d(-8)$.

   (b) Find $H(-12), H(-10), H(-8)$.

   (c) Now consider $s = a + bi$, where a and b are integers and $-15 \leq a \leq 0$ and $-2 \leq b \leq 2$. Generate a 16×5 matrix S of all possible values of s. For example, $S(1, 1)$ should be equal to $-15 - 2i$, $S(1, 2)$

should be equal to $-15 - i$, $S(1,3)$ should be equal to -15, etc. $S(2,1)$ should be equal to $-14 - 2i$, $S(2,2)$ should be equal to $-14 - i$, $S(2,3)$ should be equal to -14, etc.

(d) Find $H(s)$ for the values of s considered in (c) by performing array operations on matrix S.

(e) For what values of s considered in (c) does $H(s) = 0$? For what values does $H(s)$ blow up?

(f) Now suppose a and b are not necessarily integers. Find all the values of s for which $H(s) = 0$.

(g) Let

$$C(s) = \frac{s^2 + 2s + 2}{s^2 + 6s + 10}$$

For what values of s does $C(s)H(s) = 0$? For what values of s does $C(s)H(s)$ blow up?

3. Let $A1$ be any $n \times m$ matrix and $A2$ be any $p \times q$ matrix. Create a function **block(A1,A2)** that generates the $(n + p) \times (m + q)$ block diagonal matrix

$$\begin{bmatrix} A1 & | & 0 \\ ----- & | & ----- \\ 0 & | & A2 \end{bmatrix}$$

where the off-diagonal blocks have all elements equal to zero. Use the **zeros** command.

4. Modify the file *rms.m* so that it works properly if **v** is a row or column vector. (Note that the relational operator for equals is ==, not =.)

5. Create function **range1(p)** that returns the difference between the largest and smallest elements in the vector **p**. Use **for** and **if** statements. Now create **range2(p)** that returns the same result but use **while** and **if** statements. Is **range1** or **range2** more efficient? Modify **range1** and **range2** so that **p** can be a matrix.

2 PLOTTING

Humans are very good at extracting information from pictures. This is even recognized in our unconscious use of phrases like "I see" when we mean "I understand." There is also a cliché about the relative worth of pictures and words that we forbear repeating. MATLAB provides plotting software to help you take advantage of this ability to process visual data.

To illustrate our point about the ease with which we extract data from pictures, examine the plot shown in Figure 2.1. We have used MATLAB to create a plot of the Gaussian or normal probability density function. To demonstrate how easily we process visual information, we have deliberately added an error to the plot. You could certainly find the error by examining the 101 points that were used to produce the plot. However, it is much easier to see the error on the graph. The error is equivalent to a simple typing error—hitting the key to the right of the correct one while entering the data.

2.1 Creating Two-Dimensional Plots

Ordinary two-dimensional plots can be generated in MATLAB using the **plot** command. The most basic use of **plot** is with a single vector as the input argument. Specifically, if you enter the command **plot(v)** for any real vector **v**, MATLAB will create a plot of the elements of **v** versus their indices. That is, the first element of **v** will be plotted versus 1, the second element of **v** versus 2, etc. The result is shown in Figure 2.2

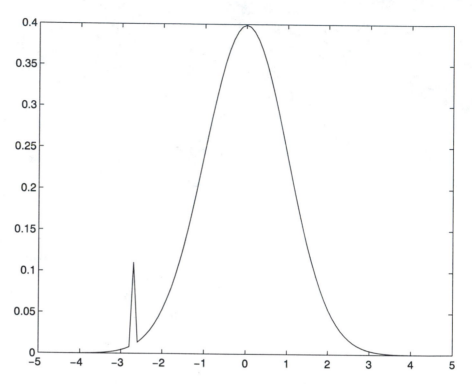

Figure 2.1   Plot of Gaussian probability density function with one erroneous point.

for **v=logspace(0,-5,12)**. Notice that the axes are automatically drawn and scaled to include all of the data in **v**. If you are working on a PC with MATLAB Version 3.5 (i.e., without Microsoft Windows) you may need to use the MATLAB commands that allow you to switch between the MATLAB command and graphics windows. When you enter a plotting command, such as **plot(v)**, in the MATLAB command window, the graphics window will automatically be called to show you your plot. You can then return to the command window by hitting any key. Subsequently, typing **shg** in the command window will switch you back to the graphics window and retrieve the most recently displayed plot.

The command **plot(v)** produces a different result if **v** is a vector of complex numbers. In this case **plot(v)** will yield a plot of the real part of **v** versus the imaginary part of **v**.

If the argument of the **plot** command is a matrix, then MATLAB treats each row as a distinct vector to be plotted and places all of the plots on the same graph. The script file *hilbplot.m* below creates a 5×5 Hilbert matrix, *H*, and plots it. The first command, **clf**, in *hilbplot.m* clears the

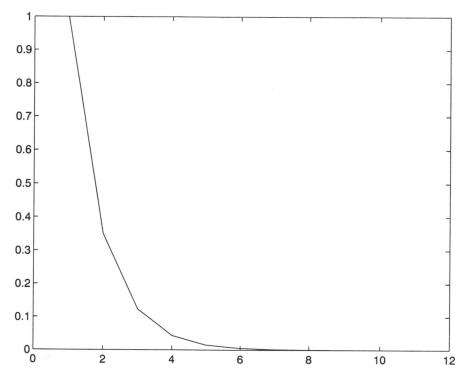

Figure 2.2   Plot of vector containing 12 elements, logarithmically spaced from 10^0 to 10^{-5}.

graphics window. It is useful to put this command in the beginning of all your plotting routines to avoid unintentionally plotting on top of a previously displayed graph. Note that **clf** replaces **clg** of MATLAB Version 3.5 although **clg** will, as of this writing, still work in Version 4.2. The result of running the file is shown in Figure 2.3.

hilbplot.m

```
% hilbplot
% Creates a 5×5 Hilbert matrix and plots its rows.
clf                    % Clear graphics window (clg in
                       % MATLAB 3.5)
H=hilb(5);             % Create a 5 by 5 Hilbert matrix
plot(H)                % Plot the matrix
```

As you can see, the result is hard to understand, primarily because the five individual plots are not labeled. We will show you how to label the lines on a plot shortly. Meanwhile, we know that the elements in each column of Hilbert's matrix are monotone decreasing. Therefore, we can

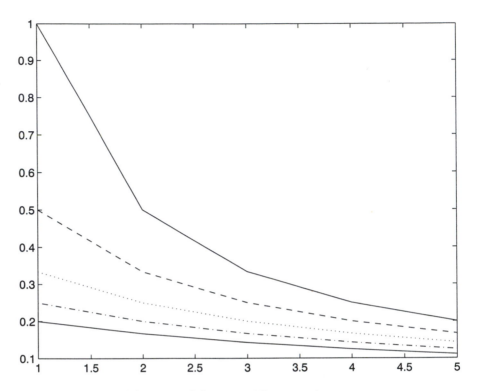

Figure 2.3 Plot of the rows of the 5× 5 Hilbert matrix.

deduce that the top plot is row 1, the next highest is row 2, etc.

If you enter **plot(x,y)** where **x** and **y** are vectors of the same dimension, MATLAB will plot **y** versus **x**. For example, Figure 2.1 was created in this manner by running the script file *gaussplot.m*, shown below.

gaussplot.m
```
% gaussplot
% Creates a plot of the Gaussian density function with
% one randomly chosen erroneous point.
clf
x=[-5:0.1:5];                % generate a vector of
                             % x-values

y=exp(-(x.^2)/2)/sqrt(2*pi);  % compute corresponding
                             % y-values
    z=10*(rand(1)-0.5);      % generate a random number
    k=1;                     % set the index to 1
    while z>=x(k)            % compare the random
                             % number to the x-values
```

```
    k=k+1;                      % increment the index while
                                % z>=x(k)
    end                         % needed termination of
                                % while loop
    y(k+1)=0.1+y(k+1);          % add error to y(k+1)
plot(x,y)                       % plot result
```

The first line of the file generates a vector of values for the x-axis (the horizontal axis) of the plot. The second line computes the corresponding y values. Notice the use of the $.\hat{\ }2$ command. This squares each component of the vector of x values. If we had used $\hat{\ }2$ we would have gotten an error message for trying to multiply two matrices of incompatible dimensions. In general, the collection of array operation commands ($.\hat{\ }$, $.*$, $./$, etc.) is very useful for plotting. If you delete the next six lines of *gaussplot.m* (these are indented in the file) leaving only the last line and run the file, then you will obtain a plot of the usual bell-shaped curve. Note that the first argument of the **plot** command is interpreted as the horizontal axis and the second argument as the vertical axis.

The extra six lines of code in *gaussplot.m* deliberately add an error to the plot. The third line in the file generates a random number z, between -5 and $+5$. The number is chosen from a uniform distribution, meaning that all numbers in the interval -5 to $+5$ are equally likely to be chosen. The work is done by the command **rand(1)**, which chooses a random number between 0 and 1 from a uniform distribution. The rest of the third line centers the random number at zero, rather than $1/2$, and scales it to the interval -5 to $+5$.

The next four lines of code compare the random number z to the elements of the vector **x**. We want to find which pair of elements of **x** bracket z. The **while** loop is a convenient way to do this. When $z<x(k)$ for the first time, the program will exit the loop. The value of **k** is the index of the largest component of **x** that is still smaller than z. We then arbitrarily add 0.1 to the next component of **y**. Figure 2.1 shows the result of running *gaussplot.m* once. As you can see, the error is extremely noticeable. If you run the file repeatedly you will get different plots each time.

For plotting multiple x versus y curves on the same graph, you have several options. For example, **plot(x1,y1,x2,y2,x3,y3)** will generate a plot with the three lines: **y1** versus **x1**, **y2** versus **x2**, and **y3** versus **x3**. Alternatively, you can use **plot(X,Y)** where **X** and **Y** are matrices. In this case, MATLAB will plot the columns of **X** versus the columns of **Y**. Enter **help plot** to learn more about these alternatives.

Once you have generated a plot, you can read data off the plot by using the command **ginput**. For example, you can use **ginput** to find the coordinates of the erroneous data point of Figure 2.1. Entering **[xerror,yerror]=ginput** will put a crosshair on the graph. You can move the crosshair with the mouse (or arrow keys if you don't have a mouse). When you have moved the crosshair to a point of interest, depress the mouse button (or any key except the enter or return key if you don't have a mouse), and MATLAB will compute the coordinates at the crosshair. You can extract coordinates for many different points at a time by successively repeating this procedure. When you are done, hit the enter or return key. You may need to be sure the crosshair is in the graphics window when you do this. MATLAB will respond in the command window by displaying the coordinates of all the data points you selected. The variable **xerror** will contain the **x** coordinates and **yerror** the **y** coordinates of the data points. There are examples of the use of **ginput** in subsequent chapters.

In addition to the **plot** command, MATLAB provides several other two-dimensional plotting commands that operate the same way as **plot** but yield different graph scales. Specifically, **loglog**, **semilogx**, and **semilogy** produce plots with one or two logarithmically scaled axes. The **polar** command generates a polar plot, and the **bar** and **stairs** commands yield bar and stairstep charts. The **bar** and **stairs** commands take only one input argument. Try using the **loglog** and other plotting commands in the previous examples by replacing **plot** with, for example, **loglog**. Also use the **help** facility for more background on these plotting commands.

MATLAB provides a function called **print** that can be used to make high-resolution hard copies of your plots. In order to use **print** for the first time, you or your system manager must edit the m-file for the function **printopt** appropriately for your type of printer. Once this is done, **printopt** need not be changed again. Entering the command **print** by itself will send the contents of the current figure directly to the printer. Using **print** followed by a file name will store the figure in a postscript file. For details on various options type **help print** and **help printopt**. If you are using Version 3.5, **printopt** will not be available, so you must edit the m-file for **print** directly. The function **print** may not be available on some computers with limited memory. Therefore, on a PC, as an alternative you can simultaneously press the shift and PrtSc keys to send a screen dump of the graphics window to the printer. On some machines with a windowing environment such as the Macintosh, you can also pull down the File menu and click on the Print command to make hard copies of your plots.

2.2 Customizing Plots

Methods for customizing plots are useful not only to add information such as titles and labels to a graph but also to better display data that are difficult to see in an ordinary, automatically scaled plot. Adding titles and labels is very straightforward as you will see; however, rescaling data to make a plot more legible may take judgment, extra work, and creativity.

The **title**, **xlabel**, and **ylabel** commands title and label the axes of your plot. Similarly, the **grid** command applies grid lines to your plot. For example, add the following four lines to the end of the file *gaussplot.m*.

title('Plot of Gaussian Probability Density Function')
xlabel('x values')
ylabel('y values')
grid

Rerun *gaussplot.m* to see how the title, labels, and grid lines appear.

Labeling of arbitrary points on a graph can be done with the **text** or **gtext** commands. The **text** command performs manual labeling; i.e., you must specify the coordinates on the graph for the position of the first letter of the label. The **gtext** command provides interactive labeling; i.e., after generating your plot enter **gtext('label')**. MATLAB will provide a crosshair in the graphics window which you can move to the desired label position using a mouse (or the arrow keys on some computers). When you are finished moving the crosshair, depress the mouse button (or a keyboard key), and MATLAB will respond by writing **label** on the plot to the right of the crosshair. Examples of **text** and **gtext** are given in the next section. Try using either or both to label the lines in the plot generated by *hilbplot.m*.

Next try replacing the command **plot(x,y)** in *gaussplot.m* with **plot(x,y,'--')** and rerun *gaussplot.m*. Instead of the default solid-line plot you should see a dashed-line plot. MATLAB provides a variety of line and point types and colors that you can use to make your plots look exactly as you wish. These are particularly useful when you need to distinguish between different lines on a multiple-line plot and you do not like the default line textures. Enter **help plot** to see a list of all the available line and point types and colors.

Using logarithmic scaling is a simple way to deal with data that contains significant features at very different orders of magnitude. The Bode plot, which is used extensively in control design, is an excellent example. It will be covered in detail in Chapter 6. An easier example is demonstrated in Figure 2.4, generated by the file *logplot.m*. In this example we also introduce the **subplot** command, which you can use to display multiple graphs simultaneously.

logplot.m

```
% logplot
% Generates a pair of plots that illustrate the
% importance of proper scaling.
clf;                          % clear graph window
                              % (clg in MATLAB 3.5)
v1=logspace(10,0,20);         % generate a logarithmic
                              % vector
v2=logspace(2,-2,20);         % generate a logarithmic
                              % vector
v=[v1 v2];                    % concatenate the two vectors
subplot(121), plot(v)         % split screen & plot the
                              % vector
subplot(122), plot(log(v)) % plot log of the vector
```

The left-hand side of Figure 2.4 shows vector **v** plotted as a function of its indices. Vector **v** consists of the concatenation of vectors **v1** and **v2**. That is, the first 20 elements of **v** are the elements of **v1**. The last 20 elements of **v** are those of **v2**. The logarithm of **v** is plotted on the right-hand side of Figure 2.4. As you can see, the values of **v2** are not visible on the left-hand (conventional) plot. The plot of the logarithm of **v** on the right clearly shows both **v1** and **v2**.

The commands **subplot(121)** and **subplot(122)** divide the graphics window into two equal halves, side by side. The **plot** command following **subplot(121)** produces a plot that is placed on the left, while the plot generated by the **plot** command following **subplot(122)** is placed on the right. In general, **subplot(abc)** (a,b,c must be integers) will create an **a**-by-**b** array of equal-sized graphs. The argument **c** indicates the current graph. Therefore, the commands **subplot(211)** and **subplot(212)** would divide the screen into two equal halves, one above the other. The syntax **subplot(a,b,c)** has become the suggested alternative starting with MATLAB Version 4.0. However, we use **subplot(abc)** in this book for compatibility with Version 3.5.

After you have used the command **subplot** you may have to input the command **subplot(111)** in order for your next plot to use the full screen.

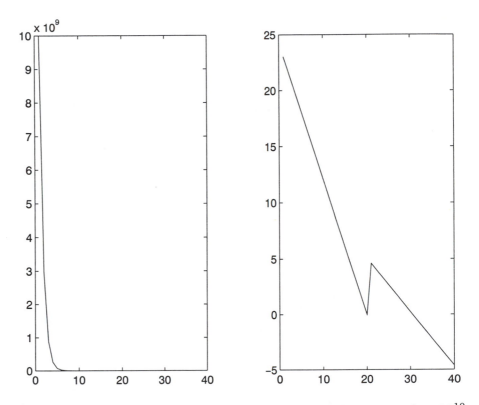

Figure 2.4 Plot of a vector containing 40 elements with values ranging from 10^{10} to 10^{-2}. Plot at left shows actual values. Plot at right shows logarithm to the base *e* of the actual values.

We have included many other examples of the use of **subplot** in this and later chapters.

As is often the case, there are other ways to produce virtually the same graphs as those in Figure 2.4. In particular, you could replace the command **plot(log(v))** by **semilogy(v)**.

It is generally a good idea to plot data in several different ways to minimize the probability that interesting features of the data are missed. Plotting the logarithm of the data is easy and often improves the scaling. Unfortunately, there are many situations where other methods are needed. Remember that only positive numbers have real logarithms. Also, distorting the scale of a graph with logarithmic scaling may be inappropriate or misleading. Because the logarithm compresses large numbers and makes small positive numbers into large negative ones, it can, for example, turn a plot of $1.01+\sin(t)$ versus t, which you would immediately recognize, into something unrecognizable.

The following example illustrates another way of dealing with poorly scaled data. A small signal (sinuosidal in this case) superimposed on a large one (a square wave for simplicity) is common in engineering and illustrates the difficulty of displaying small signals on a large scale. Electrical engineers often find small amounts of 60 Hz ripple due to the power lines superimposed on other signals. File *ripple.m* demonstrates a technique for displaying both portions of the combined signal.

ripple.m
```
% ripple
% Creates plots that illustrate the value of removing
% large known signals from the plot.
clf
vwithripple=[100*ones(1,20) zeros(1,20)]+0.1*sin(1:40);
plot(vwithripple)
keyboard
subplot(121),        plot(vwithripple(1:20))
subplot(122),        plot((21:40),vwithripple(21:40))
```

The fifth line of the file creates one cycle of a square wave of amplitude 100 and period 40 plus a ripple signal $0.1\sin(t)$. The sixth line produces the plot shown in Figure 2.5. The sinusoidal ripple is invisible, as you can see. The problem is that the ripple signal is smaller than the thickness of the line corresponding to the square wave at the scale needed to display the square wave. The **keyboard** command in the file causes MATLAB to interrupt running of the script file and return control to the command window. If we were to delete the **keyboard** command, the next line of *ripple.m* would cause Figure 2.5 to be overwritten. By including the **keyboard** command we are given the opportunity to make a hard copy of the plot, e.g., with the **print** command. Typing the word **return** and then striking the "return" or "enter" key will restore running of the script file *ripple.m*.

The last two lines of the file create side-by-side plots of the two halves of the signal. These are displayed in Figure 2.6. The first half of the cycle, when the square wave signal is constant at 100, shows up on the left-hand-side plot and the second half of the cycle, when the square wave signal is constant at 0, shows up on the right-hand-side plot. The ripple is quite evident in these plots. Because the square wave is constant in each subplot, MATLAB automatically scales the display to show the ripple, the only varying signal. The figure needs to be examined carefully to see that the vertical axis scales are very different in the two plots of Figure 2.6.

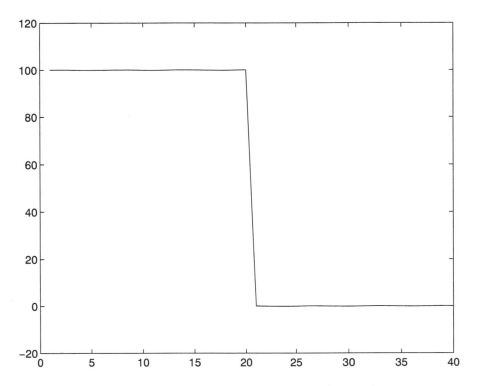

Figure 2.5   Plot of one cycle of a square wave plus ripple signal.

Notice the somewhat denser coding of the file *ripple.m* compared with our previous examples. This is slightly more efficient in the sense that it uses slightly less memory and runs a little faster. However, it is harder to understand the code and, as a result, it took longer to debug. We will generally try to keep our examples readable. Many of them could be made more efficient by coding techniques similar to those used in *ripple.m*.

Two other MATLAB commands, **hold** and **axis**, can be used for customizing graphs. However, these commands need to be used carefully because they can introduce problems if they are used incorrectly. The **hold** command is used for superimposing one graph on another. Specifically, **hold on** maintains the current plot in the MATLAB graphics window and all subsequent plotting commands are superimposed on the held plot until **hold off** is entered. The **hold** command toggles between **hold on** and **hold off**. Because you may not know the current hold state, using **hold on** and **hold off** is preferable. However, you can determine the current hold state using the function **ishold**, which returns **1** if **hold** is on and **0** if it is off.

The **axis** command can be used to select manually the ranges of the plot

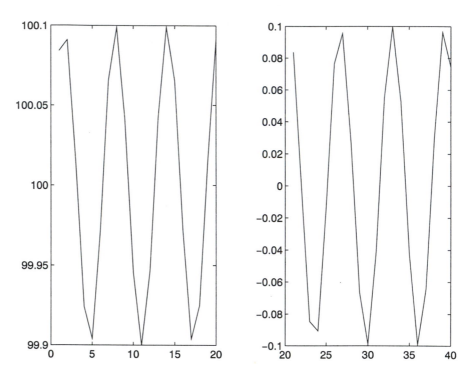

Figure 2.6 Blown-up plots showing the ripple of the two halves of the signal shown in Figure 2.5.

scales. For example, **axis([–5,5,0,10])** will set the horizontal axis to range from –5 to 5 and the vertical axis from 0 to 10. Typing **hold on** and then **axis(axis)** on the next line will freeze the current scale. Typing **axis ('auto')** will return MATLAB to automatic scaling. Be careful when using **axis** that you do not freeze a scale when you actually want automatic scaling and vice versa. Use of the **axis** command is demonstrated in later chapters.

2.3 Plotting Three-Dimensional Data

MATLAB enables you to display three-dimensional information in three ways: contour plots, mesh plots, and plots of lines in three-dimensional space. Plots of lines are done using the command **plot3**, which is the three-dimensional analog of the **plot** command. Contour and mesh plots are demonstrated in file *simple3d.m*, which generated the plots shown in Figure 2.7 using the **contour** and **mesh** commands. The expression that

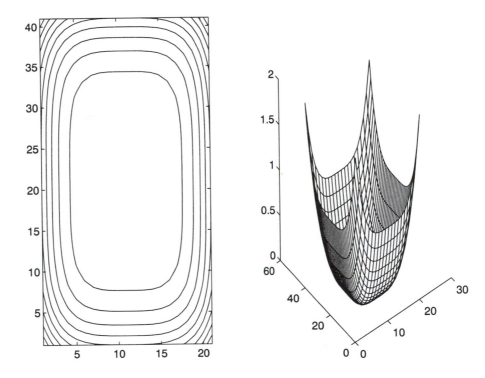

Figure 2.7   Contour and mesh plots of $z = x^4 + (y/2)^4$.

is plotted is $z = x^4 + (y/2)^4$. The "vertical" axis of the plots is **z**, which represents z. The first two lines of the code in the file create vectors **x** and **y** (i.e., x and y). These will be the axes of the "horizontal" plane. The third line introduces the command **meshgrid**. This replaces the command **meshdom** from MATLAB Version 3.5. This command takes the vectors **x** and **y** and creates the arrays **X** and **Y**, which can be used to evaluate a function of two variables such as **z**. Each row of **X** is equal to **x**, and the number of rows of **X** is equal to the length of **y**. Each column of **Y** is **y'**, and the number of columns of **Y** is equal to the length of **x**. The next line computes **z** as a matrix by performing array operations on **X** and **Y**. There is a point in matrix **z** that corresponds to each point in the rectangular grid produced by **meshgrid**. The last two lines produce a contour plot and a 3D mesh plot.

simple3d.m

```
% simple3d
% This creates a contour plot of a simple quartic
% function on the left of the screen and a 3D
% perspective (mesh) plot of the same function on
```

```
% the right.
clf
x=[-1:.1:1];
y=[-2:.1:2];
[X,Y]=meshgrid(x,y);
z=X.^4+(Y/2).^4;
subplot(121), contour(z)
subplot(122), mesh(z)
```

Each curve in the contour plot represents the values of **x** and **y** that correspond to a fixed value of **z**. The x-y (horizontal) plane coincides with the page. The "vertical" z-axis comes directly out of the paper. Although it is difficult to tell by inspection, as the values of **z** are increased the curves enclose larger and larger areas. Notice that each contour seems like the hybrid offspring of a circle and a rectangle. This is the effect of using the fourth power of **x** and **y** rather than the second power, which would have produced elliptic contours. The mesh plot gives a tilted 3D perspective rendering of the same data. It is a good idea to display both the contour and mesh plots. The mesh plot helps you visualize the 3D shape and assign heights to the contours. It is also possible to label the contours in the contour plot using the command **clabel**.

Note that the axes of both the contour and mesh plots are inaccurately labeled. Unless you specify otherwise, MATLAB assumes that the horizontal axes are the indices of the elements of the matrix **z**. This is easily fixed by replacing **mesh(z)** by **mesh(x,y,z)** and **contour(z)** by **contour(x,y,z,10)** in file *simple3d.m*. In the **mesh** command **x** and **y** specify the horizontal axes and **z** specifies the vertical axis. In the **contour** command the number 10 specifies the number of individual contour lines to be plotted while the **x** and **y** specify the horizontal and vertical axes, respectively. The ordering of the variables differs in MATLAB 3.5. Use the help facility for the appropriate syntax.

Thus far, in order to emphasize the basics of plotting in MATLAB, we have been deficient about labeling our graphs. As a general rule, at least the axes of graphs should be carefully labeled. It is also desirable to use the **text** and **gtext** commands described earlier to label the individual curves of any graph containing multiple curves. Such labeling is used in the remaining examples.

The problems of poorly scaled data also arise in three dimensions. The Rosenbrock banana function is a good illustration of what can happen with such data. The function is $h(x_1, x_2) = 100(x_2 - x_1^2)^2 + (1 - x_1)^2$.

You can see immediately that this function has a minimum at $x_1 = x_2 = 1$. It is not immediately obvious why it is called the Rosenbrock banana function. Howard Rosenbrock proposed this function as a difficult test for algorithms for minimizing functions. The name derives from the fact that this particular function, viewed in 3D, has a steep banana-curved valley around its minimum. The steepness and curvature cause simple, gradient-based, function-minimization algorithms to fail to find the minimum.

We have created the function **banana** both to illustrate the creation of functions and to facilitate plotting of the Rosenbrock banana function. The file *banana.m* creates the function **banana** that computes **h** given **x1** and **x2** according to the equation above.

banana.m
```
function h=banana (x1,x2)
% function banana is the Rosenbrock banana function
% usage:   h=banana(x1,x2)
h=100*(x2-x1.^2).^2 + (ones(size(x1))-x1).^2;
```

Using the function **banana**, we then created both contour and mesh plots of the function as shown in Figure 2.8. The plots were created by running file *bananam1.m*; **x** (equivalent to x_1) was allowed to range from 0 to 2, and **y** (equivalent to x_2) was allowed to range from 0 to 4. The function **banana** is used to compute $h(x_1, x_2)$ and store the results in **z**.

bananam1.m
```
% bananam1
% This creates a contour plot of the Rosenbrock banana
% function on the left and a 3D plot of the same
% function on the right.
clf
x=[0:.1:2];
y=[0:.1:4];
[X,Y]=meshgrid(x,y);   % Use meshdom if you have
                       % MATLAB 3.5
z=banana(X,Y);
subplot(121), contour(x,y,z)
subplot(122), mesh(x,y,z)
```

As you can see in Figure 2.8, it is very hard to see any trace of the steeply curving valley that gives the function its name. To convince you that the

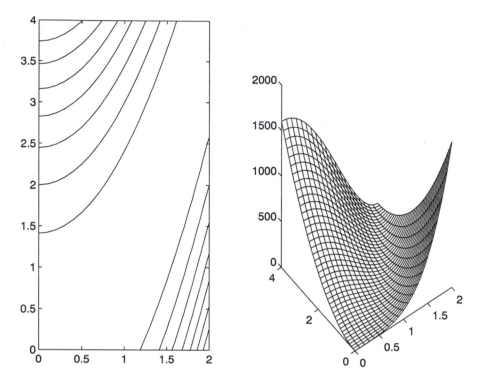

Figure 2.8 Contour and mesh plots of the Rosenbrock banana function.

valley is actually there and to demonstrate one way of seeing it, we created file *bananam2.m*, which plots the logarithm of the banana function. Running *bananam2.m* produces Figure 2.9. As you can see, plotting the logarithm of the function illustrates much more clearly the feature that gives the banana function its name. Note also our use of the MATLAB Version 4.2 command **meshc** and the resulting contour plot below the mesh plot.

bananam2.m

```
% bananam2
% This creates a plot of the logarithm of the Rosenbrock
% banana function.
clf
x=[0:.1:2];
y=[0:.1:4];
[X,Y]=meshgrid(x,y);
z=banana(X,Y);
logz=log(1+z);    % Note the use of z+1 to avoid log(0)
meshc(x,y,logz)   % meshc is not available in MATLAB 3.5
```

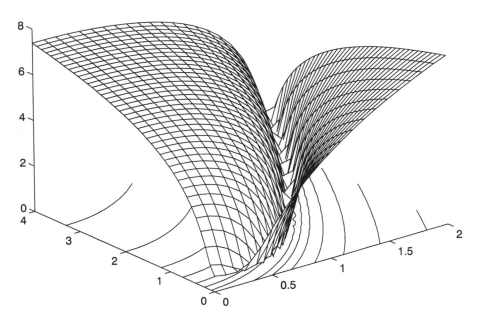

Figure 2.9 Mesh plot and contour plot of the logarithm of one plus the Rosen-brock banana function.

Another approach for illustrating the banana function using a contour plot is illustrated in file *bananacontour.m* below.

bananacontour.m

```
% bananacontour
% This creates a contour plot of the Rosenbrock banana
% function.
clf
x=[0:0.005:2];
y=[0:0.005:4];
[xmesh,ymesh]=meshgrid(x,y);
z=banana(xmesh,ymesh);
v=logspace(-2,0,5);
contour(x,y,z,v)
xlabel('x1')
ylabel('x2')
gtext('Contour Plot of Rosenbrock banana function')
```

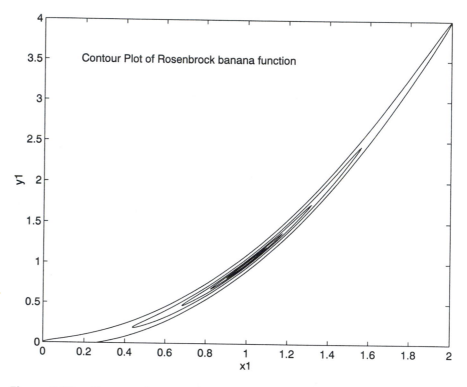

Figure 2.10   Contour plot using logarithmically spaced contours.

The crucial step in this file is the creation and use of the vector **v** to control which contours are plotted. We plot five curves for values of **z** between 0.01 and 1.0. These values of **z** are logarithmically spaced. The result is shown in Figure 2.10. Note that the mesh size in *bananacontour.m* is much smaller than in *bananam2.m*. Mesh plots seem to be clearer with fairly coarse mesh sizes. However, the banana function needs a fine mesh in order for **z** to be accurate enough for a good contour plot. You can easily verify this by running *bananacontour.m* yourself using a coarser mesh. In fact, you may have some trouble running *bananacontour.m* on a small machine unless you make the mesh coarser. The result is actually quite interesting.

Notice that Figure 2.10 is relatively well labeled. This is accomplished by the last three lines of *bananacontour.m*, which illustrate the use of the commands **xlabel**, **ylabel**, and **gtext**. Notice that the text for each of these commands must be enclosed between single closed quotation marks and not double quotation marks. Also, **gtext** creates a crosshair (when you move the pointer into the graphics window). Click the mouse to insert the titles.

Our final example in this chapter is designed to illustrate why poles and zeros are called poles and zeros. As poles and zeros are not formally introduced until Chapter 4, you may want to reread this example after reading Chapter 4. In this example we also demonstrate another way to deal with badly scaled data and some more techniques for labeling graphs. The example is in file *polesandzeros.m*.

polesandzeros.m

```
% polesandzeros
% This creates a contour plot above and a mesh plot
% below for G(s)=(s^2+3s+2)/(s^2+2s+2)(s+4) and
% labels the plot.
clf
num=[1 3 2];
den=conv([1 2 2], [1 4]);
x=[-5:0.1:0];
y=[-2:0.1:2];
X=ones(size(y'))*x;
Y=y'*ones(size(x));
s=X+Y*i;
numval=polyval(num,s);
denval=polyval(den,s);
temp3=numval./denval;
N=abs(temp3);
sz=size(temp3);
% The following loops chop off the peaks of the 3 poles.
% isinf and isnan are not available in MATLAB 3.5.
for l=1:sz(1)
    for k=1:sz(2)
      if ((N(l,k)>5.0) | isinf (N(l,k)) | isnan(N(l,k)),
        N(l,k)=5.0;
      end
    end
end
v=[0.05:0.1:2];
subplot(211),
contour(x,y,N,v)
text(-1.91,-0.1,'.05')
text(-0.97,-0.1,'.05')
text(-2.83,0.15,'.25')
text(-3.25,0.5,'.45')
```

```
text(-1.68,1.4,'.45')
text(-1.68,-1.4,'.45')
xlabel('Real')
ylabel('Imaginary')
subplot(212),
meshz(x,y,N)
% note the use of meshz (MATLAB 4.2)
text(-3.5,-3.8,'Real')
text(-7.5,-1,'Imaginary')
```

Running the file *polesandzeros.m* produces contour and mesh plots of the magnitude of the transfer function $G(s) = (s^2 + 3s + 2)/(s^2 + 2s + 2)(s+4)$ on the complex plane as shown in Figure 2.11. Detailed instructions for specifying transfer functions are provided in Chapter 3; however, you should note that the second and third lines of the file create numerator and denominator polynomials **num** and **den** for $G(s)$. Once we take **num/den** we will have a transfer function with zeros at −1 and −2 and poles at $-1 + i, -1 - i$, and -4. The fourth and fifth lines create real (x) and imaginary (y) axis vectors. The next two lines create a mesh spanning a portion of the $x - y$ plane. These two lines could be replaced by a single line using **meshgrid**. The next four lines create a matrix, called **temp3**, that contains the values of our transfer function at each point of the mesh.

Note that **temp3** is a matrix of complex numbers. The next line uses **abs** to compute the matrix **N**, the elements of which are the moduli of the complex numbers in **temp3**. However, **temp3** contains several **NaN's** and **Inf's**. After all, our transfer function has three poles and should blow up at each of them. The **for** loops take care of the problem by setting all elements of **N** that are **NaN, Inf**, or >5 to be equal to 5. Notice that this method, simply sawing off the peaks, is another way to deal with badly scaled data.

Another interesting feature of file *polesandzeros.m* is the use of the text command to label the plot. It is particularly important to label the curves on the contour plot; otherwise, the zeros look quite similar to the poles. Notice also that it is necessary to use **text** to label the axes of the mesh plot. Finally, do not think that the first two arguments of the **text** commands, which position the text on the graph, were arrived at without some experimentation. It took us several attempts before we were able to place the labels properly.

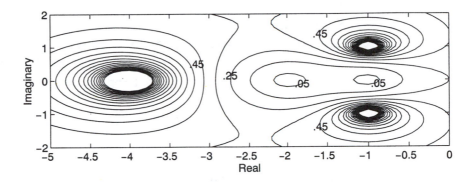

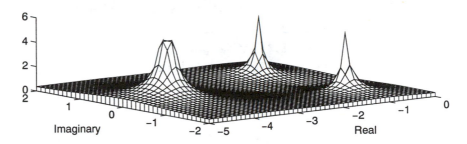

Figure 2.11   Contour and mesh plots of $|G(s)| = |(s^2 + 3s + 2)/(s^2 + 2s + 2)(s + 4)|$.

2.4 Introduction to Handle Graphics

It is often desirable to be able to customize the graphical output from MATLAB or any other software package. We have already demonstrated how to use some of the higher-level MATLAB commands such as **xlabel**, **ylabel**, **text**, and **gtext** to customize graphs. MATLAB Version 4.2 provides a large collection of lower-level commands, referred to as Handle Graphics, that gives the user control over the details of MATLAB-produced plots.

The m-file *introhan.m* illustrates some of the Handle Graphics capabilities and their use. The first three lines of code are a duplicate of *hilbplot.m*. If you have already created and stored *hilbplot.m*, you could replace them by the command **hilbplot**. If you do this and you do not get the plot on your screen, then you have to correct MATLABPATH so that MATLAB can find *hilbplot.m*.

introhan.m

```
% introhan
% An introduction to handle graphics.
% Creates a 5×5 Hilbert matrix and plots its rows.
clf                         % Clears the graphics window
H=hilb(5);                  % Create a 5×5 Hilbert matrix
plot(H)                     % Plot the matrix
h=gcf;
h1=get(h,'CurrentAxes');
set(h1,'LineWidth',3)
h2=get(h1,'Title');
set(h2,'String','Plot of the rows of the Hilbert matrix')
set(h2,'FontSize',14)
set(h2,'FontWeight','bold')
set(h2,'FontName','Courier')
h3=get(h1,'Children');
set(h3,'LineWidth',2)
```

The next three lines change the thickness of the lines that form the axes of the plot. The following five lines (all five include **h2** somewhere in them) create a title for the plot and display that title using the Courier font in bold 14-point letters. The MATLAB default is Helvetica in normal weight 12-point letters. The last two lines of the m-file change the thickness of the five plotted lines.

The principle behind *introhan.m* is that certain features of MATLAB plots are assigned numbers (represented by the variables **h**, **h1**, **h2**, and **h3** in our example) and can be accessed by means of those numbers. The command **gcf** gets the number of the current figure. The command **set(h)** lists every feature directly associated with **h** and its possible settings. The command **get(h)** lists the same features and their current settings. (Try typing these commands.) You can then change some of these settings. For example, the command **set(h,'NumberTitle','off')** will remove the title Figure No. 1 from the display created by *introhan.m*.

The feature **CurrentAxes** is a property of the current figure and has a number assigned to it by MATLAB. In *introhan.m* we store that number as **h1**. Note that it is obtained by **h1=get(h,'CurrentAxes')**. Alternatively, **h1=gca** would work. It is a good idea to make this variable assignment rather than to display the number (i.e., leave out the **h1=**) and try to retype it when accessing the axes. The actual number is a floating point number that must be entered exactly when accessing the axes. MATLAB's default display rounds to six significant figures. Thus, reading the display and retyping the number can lead to errors even when the displayed number is retyped correctly.

Once you have defined **h1**, the commands **set(h1)** and **get(h1)** work as they did for **h**. Settable features are changed as **'NumberTitle'** was above. We have illustrated changing the thickness of the current axes in the sixth line of code of *introhan.m*.

CurrentAxes has several features that are also assigned numbers by MAT-LAB. We obtain them and store them in **h2** and **h3**. Their associated features are modified in the same way as those of **CurrentAxes**. This is illustrated in the remaining lines of *introhan.m*.

Handle Graphics provides many other possibilities for improving the appearance of graphical output, creating graphs, and even creating interactive software in which the user interface includes push buttons and menus. None of this is particularly hard to learn or to do once you get started. These capabilities are not central to the use of MATLAB for controls, so we leave further exploration of them to the reader.

2.5 Exercises

1. Create **bar**, **stairs**, **semilogy**, and **loglog** plots of the Gaussian probability density function.

2. Replace the **plot(log(v))** command in file *logplot.m* by **semilogy(v)**, run the new file, and compare the result with Figure 2.4.

3. Create and run the following file.

   hiddensine.m

   ```
   v=(100*rand(1))*sin((rand(1)*10)*[0:0.1:10])+...
        0.1*sin([0:0.1:10])

   plot(v)
   ```

   Write a script file that will identify and cancel the random signal and plot the small sinusoid. Your solution can use only the vector **v** and the plot.

4. Create mesh and contour plots of the Rosenbrock banana function on the region $-2 \le x \le 2$, $-3 \le y \le 3$. Is this a better choice of region to

plot than the one in Section 2.3?

5. Extract $|G(s)|$ for $s = j\omega$ ($0 \leq \omega \leq 2$) from the data produced by file *polesandzeros.m*. Plot $|G(s)|$ vs. ω for the above values of ω.

6. If you have MATLAB 4.2, rerun *polesandzeros.m* using **surf** and **waterfall** in place of **meshz**.

3 SETTING UP CONTROL PROBLEMS

A deep understanding of the connections among the different descriptions of linear time-invariant systems is essential to the design of single-input, single-output (SISO) control systems. Control system designers as well as theoreticians switch back and forth among transfer function, state-space, pole-zero, and impulse response descriptions of linear time-invariant systems in their efforts to create and understand controls. While each description has its advantages and disadvantages, each one is important because it describes different facets of the behavior of the system.

MATLAB provides facilities for defining a system according to any one of these descriptions and switching from one description to another. It also provides facilities for graphical display of the different descriptions. The primary emphasis of the first few chapters of this book was the basics of MATLAB itself. The rest of the book, beginning with this chapter, deals primarily with the use of MATLAB and its Control System Toolbox for the analysis and design of linear time-invariant SISO control systems. This chapter shows you how to use MATLAB to set up transfer function and state-space descriptions of whatever system is of interest.

3.1 Creating Transfer Functions

Consider the system described by

$$\frac{Y(s)}{U(s)} = G(s) = \frac{18(s + 20)}{(s + 15)(s + 25)(s + 0.4)}$$

where the transfer function $G(s)$ relates the output $Y(s)$ to the input $U(s)$. This particular transfer function example relates the propeller shaft speed $Y(s)$ to the fuel input $U(s)$ for a gas turbine-powered marine propulsion plant. The script file *marineplant.m* creates the transfer function description of this system in MATLAB.

marineplant.m

```
% marineplant
% This creates, multiplies out, and prints
% the transfer function.
% G(s)=18(S+20)/(S+15)(S+25)(S+0.4)
num=18*[1 20];
den=conv(conv([1 15], [1 25]), [1 0.4]);
printsys(num,den,'s')
```

The variable **num** represents the numerator polynomial of the transfer function $G(s)$. As discussed in Chapter 1, polynomials are described by row vectors with elements equal to the polynomial coefficients in order of decreasing powers. Similarly, **den** represents the denominator polynomial of the transfer function. The **printsys** command as given in *marineplant.m* writes out the transfer function equivalent to $num(s)/den(s)$. The last argument **'s'** indicates that the transfer function should be written in the *s*-domain. Entering **marineplant** yields

```
           18 s + 360
   - - - - - - - - - - - - - - - - - - -
   s^3 + 40.4 s^2 + 391 s +150
```

A mass, spring, and damper system has second-order dynamics of the form

$$\frac{X(s)}{F(s)} = G_m(s) = \frac{1}{Ms^2 + \mu s + K}$$

where $X(s)$ is the position of the mass M, $F(s)$ is the force applied to the mass, μ is the damping coefficient and K is the spring stiffness. More complicated systems can sometimes also be approximated by second-order dynamics. Second-order systems are commonly written in the standardized form

$$\frac{Y(s)}{U(s)} = G_2(s) = \frac{k}{s^2 + 2\zeta\omega_n s + \omega_n^2}$$

where ω_n represents the natural frequency and ζ the normalized damping ratio of the system. When we transform the mass, spring, and damper system dynamics into this form we get $k = 1/M$, $\omega_n = \sqrt{K/M}$, and $\zeta = \mu/2\sqrt{KM}$.

Because it is so common to describe systems in this form, MATLAB provides the command **ord2** to generate a system description given values for ζ and ω_n. For example, the file *masspringdamp.m* contains the commands to generate the transfer function description for a second-order system with $k = 1$, $\omega_n = 2$, and $\zeta = 0.707$ (e.g., $M = 1$, $K = 4$, $\mu = 2.828$ for the mass, spring, damper system).

masspringdamp.m

```
% masspringdamp
% Illustration of the use of ord2.
wn=2;
damping=0.707;
[num2,den2]=ord2(wn,damping);
printsys(num2,den2,'s')
```

The **ord2** command uses two input arguments, natural frequency and damping ratio, and provides two output arguments corresponding to the numerator and denominator of the equivalent transfer function. MATLAB automatically sets $k = 1$. Note that the output arguments, called **num2** and **den2** in this example, must be placed between brackets. Entering **masspringdamp** at the MATLAB prompt produces

```
num/den =
                    1
        - - - - - - - - - - - -
            s^2 + 2.828 s+4
```

MATLAB also provides a means of generating a random transfer function. The command **[numr,denr]=rmodel(n)** produces the numerator **numr** and denominator **denr** of a randomly generated stable transfer function of **n**th order.

Finally, suppose your system includes a time delay that is T seconds long. The Laplace transform for such a delay is given by e^{-sT}. MATLAB does not provide an exact means of describing such a delay. However, the **pade** command can be used to generate a transfer function that approximates the time delay. Specifically, the command **[num,den]=pade(T,n)** will generate the numerator and denominator of an **n**th order Pade approximation of e^{-sT}.

Once you have generated a transfer function description in MATLAB, there is a variety of tools you can use to study and analyze the system. For example, you can plot step responses using **step**, root locus plots with **rlocus**, and Bode plots with **bode**. These and many more tools will be described in subsequent chapters.

3.2 Creating State-Space Models

Creating state space models in MATLAB is analogous to creating transfer function descriptions. Consider the standard linear time-invariant SISO state-space model description

$$\dot{x}(t) = ax(t) + bu(t)$$

$$y(t) = cx(t) + du(t)$$

where vector $x(t)$ represents the state, scalar $u(t)$ the input, and scalar $y(t)$ the output of the system. The constant a is a matrix, b a column vector, c a row vector, and d a scalar. Since MATLAB makes matrix and vector generation and manipulation easy, setting up a state-space model is very straightforward. The file *ssmarineplant.m* creates a state-space model of the marine propulsion plant described by $G(s)$ in Section 3.1.

ssmarineplant.m
```
% ssmarineplant
% State space version of the marine propulsion plant.
a=[-40.4 -391 -150; 1 0 0; 0 1 0];
b=[1 0 0]';
c=[0 18 360];
d=0;
printsys(a,b,c,d)
```

The constants **a**, **b**, **c**, and **d** are generated using the standard MATLAB notation for matrices and vectors. The command **printsys** in this example writes out the state-space model as provided. Entering **ssmarineplant** at the MATLAB prompt yields:

```
a =
              x1            x2            x3
     x1       -40.40000     -391.00000    -150.00000
     x2       1.00000       0             0
     x3       0             1.00000       0
b =
              u1
```

```
        x1              1.00000
        x2              0
        x3              0
c  =
                        x1              x2              x3
        y1              0               18.00000        360.00000
d  =
                        u1
        y1              0
```

Similarly, the **ord2** command can be used to generate a state-space model for a second-order system given the system natural frequency and damping ratio. We simply need to define the output arguments of the **ord2** command in terms of the state-space model constants instead of the transfer function numerator and denominator. File *ssmassspringdamp.m* generates the state-space model for the example given in Section 3.1 where natural frequency $\omega_n = 2$ and damping $\zeta = 0.707$.

ssmassspringdamp.m
```
% ssmassspringdamp
% Illustration of the state space version of ord2.
wn=2;
damping=0.707;
[a2,b2,c2,d2]=ord2(wn,damping);
printsys(a2,b2,c2,d2)
```

Running this file produces

```
    a  =
                        x1              x2
            x1          0               1.00000
            x2          -4.00000        -2.82800
    b  =
                            u1
            x1          0
            x2          1.00000
    c  =
                        x1              x2
            y1          1.00000         0
    d  =
                        u1
            y1          0
```

The **rmodel** and **pade** commands can also be used to generate random, stable, state-space models and state-space Pade approximations to time delays, respectively. Additionally, the **canon** command can be used to transform an arbitrary state-space model into its canonical form, discussed further in Chapter 7. The **minreal** command produces a new state-space model from an arbitrary one, eliminating states that are unnecessary or redundant. The **minreal** command can also be used to modify transfer function descriptions. State-space model order reduction is also performed with the **modred** command. We provide examples of **modred** and **balreal** (which creates a balanced state-space realization) in Chapter 7.

As with the transfer function descriptions, once the state-space models are input into MATLAB they can be studied and analyzed extensively as described in the remainder of this book.

3.3 Changing from State Space to Transfer Function

As mentioned at the beginning of this chapter, control engineers need to be adept at switching among the various system descriptions. MATLAB makes switching easy. For example, the command **ss2tf** switches from state space (ss) to (2) a transfer function (tf). Note that if you learn these mnemonics you will easily be able to figure out other similar commands, such as **tf2ss**, which switches from a transfer function (tf) to (2) a state-space (ss) model.

After having run *ssmasspringdamp.m*, which generates the state-space model **a2**, **b2**, **c2**, **d2**, try running *convertssmass.m*, which converts the state-space model generated by *ssmasspringdamp.m* into a transfer function.

convertssmass.m

```
% convertssmass
% This converts the second-order state space system
% created earlier into transfer function form
% (assumes a2, etc. already present in workspace).
[numcss,dencss]=ss2tf(a2,b2,c2,d2,1);
printsys(numcss,dencss,'s')
```

MATLAB will respond with

num/den =

$$\frac{1}{s\hat{\ }2 + 2.828s + 4}$$

which is identical to the original transfer function description of the mass, spring, damper system generated with the file *masspringdamp.m.* The last input argument in the **ss2tf** command indicates which input of the state-space model is of interest. For SISO systems this argument should always be 1 or it can be omitted. However, for multi-input systems other values may be selected.

You will find that **ss2tf** works reliably for simple, low-order systems. However, **ss2tf** uses a numerically unstable algorithm. To determine the transfer function it first converts the state-space model into the companion canonical form. Unfortunately, the companion canonical form can be poorly conditioned. As a result, you must be very careful using **ss2tf**.

Additionally, transfer function descriptions themselves can often be ill-conditioned. This comes about because transfer functions are described as the quotient of two polynomials, which are in turn described by their coefficients. Roots of polynomials are very sensitive to the polynomial coefficients. Slight errors in the coefficients can dramatically alter the roots. Therefore, slight numerical errors introduced by **ss2tf** into the polynomial coefficients can greatly influence important features of the computed transfer function. We provide an example of these conditioning problems in the next section, where we introduce **tf2ss** to switch from a transfer function back to state space.

3.4 Changing from Transfer Function to State Space

Just as we can convert a state-space model into a transfer function de-scription, we can use **tf2ss** to perform the reverse conversion. Suppose you have created the second-order transfer function generated by run-ning the file *masspringdamp.m.* To convert this transfer function into a state-space model run the file *converttfmass.m.*

converttfmass.m
% converttfmass
% This converts the mass, spring, dashpot model back to state space.

```
[actf,bctf,cctf,dctf] = tf2ss(numcss,dencss);
printsys(actf,bctf,cctf,dctf)
```

MATLAB will respond with the state-space model

```
        a =
                        x1              x2
                x1      -2.82800        -4.00000
                x2       1.00000        0
        b =
                        u1
                x1      1.00000
                x2      0
        c =
                        x1              x2
                y1      0               1.00000
        d =
                        u1
                y1      0
```

Switching between system descriptions will not always be as exact as it is for the simple second-order system just illustrated. This is mostly due to the inherent poor conditioning of transfer functions and the poor conditioning of companion canonical form state-space models. MATLAB functions, such as **ss2tf** and **tf2ss**, that involve operations or conversions of transfer function descriptions can create errors by magnifying small inaccuracies or round-off errors. Transfer functions with repeated roots in the denominator tend to be particularly difficult to manipulate precisely.

For example, consider the transfer function

$$G_{ill}(s) = \frac{1}{(100s^2 + s + 1)^2}$$

which has a denominator with two sets of roots equal to $-0.005 \pm 0.1i$. Because there are multiple roots in the denominator and the coefficients of the denominator polynomial vary greatly in magnitude, this transfer function is ill-conditioned. To illustrate this point we convert $G_{ill}(s)$ to a state-space model and then back to a transfer function description in the file *backandforth.m*.

backandforth.m
```
% backandforth
% This demonstrates an ill-conditioned transfer
% function.
```

```
numill=1;
denill=conv([100 1 1], [100 1 1]);
printsys(numill,denill,'s')
[aback,bback,cback,dback]=tf2ss(numill,denill);
[numforth,denforth]=ss2tf(aback,bback,cback,dback);
printsys(numforth,denforth,'s')
condition_number=cond(aback)
```

If you enter **backandforth**, MATLAB will respond by printing the original transfer function description, the new transfer function (i.e., the result of the back-and-forth conversions), as well as the condition number of the matrix **aback** from the intermediate state-space model, as follows:

num/den =

$$\frac{1}{1e+04 \ s\hat{\ }4 + 200 \ s\hat{\ }3 + 201 \ s\hat{\ }2 + 2 \ s+1}$$

num/den =

$$\frac{6.939e-18 \ s\hat{\ }3 - 1.041e-17 \ s\hat{\ }2 + 2.277e-18 \ s+0.0001}{s\hat{\ }4 + 0.02 \ s\hat{\ }3 + 0.0201 \ s\hat{\ }2 + 0.0002 \ s + 0.0001}$$

condition_number =
        1.0008e+04

Notice that the new transfer function is scaled by 10^4 and has additional terms in the numerator. Additionally, the condition number of **aback** is exceedingly high. The errors in the numerator of the new transfer function are actually quite small ($10^{-17} \approx 0$). Therefore, we could recover the original transfer function by the obvious and reasonable method of setting very small coefficients to zero. However, since **aback** is so poorly conditioned, we could run into trouble if we needed to use the intermediate state-space model for further calculations.

One way to reduce these errors, in particular the condition number of **aback**, is to treat the fourth-order system described by $G_{\text{ill}}(s)$ as two second-order systems in series:

$$G_{\text{ill}}(s) = \left(\frac{1}{100s^2 + s + 1} \right) \left(\frac{1}{100s^2 + s + 1} \right) = G_1(s)G_1(s)$$

Instead of converting $G_{\text{ill}}(s)$ into a state-space model in one step, we convert $G_1(s)$ and then use the **series** command to connect the resulting state-space model in series with itself in the file *seriesconvert.m*.

seriesconvert.m

```
% seriesconvert
% This demonstrates one way to improve the condition
% number.
numg1=1;
deng1=[100 1 1];
[ag1,bg1,cg1,dg1]=tf2ss(numg1,deng1);
[ag,bg,cg,dg]=series(ag1,bg1,cg1,dg1,ag1,bg1,cg1,dg1);
[numg,deng]=ss2tf(ag,bg,cg,dg);
condition_number=cond(ag)
```

MATLAB will respond to **seriesconvert** with

```
condition_number =
        161.8196
```

That is, the condition number of the state-space representation of $G_{ill}(s)$ has been reduced by a factor of 100.

The MATLAB command **series** provides the series connection of the systems **ag1, bg1, cg1, dg1** and **ag1, bg1, cg1, dg1** in *seriesconvert.m.* The input arguments for **series** are the system descriptions for each of the two systems being connected. The output arguments describe the connected system. **series** can be used analogously for transfer function descriptions. Similarly, the MATLAB command **parallel** generates the overall system description for two systems connected in parallel. For example, to describe the system of Figure 3.1 with a transfer function, $H(s)$, execute file *parallelex.m.*

parallelex.m

```
% parallelex
% Illustration of parallel connection of systems.
```

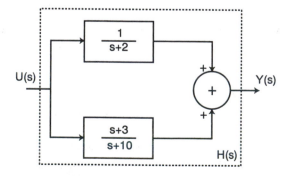

Figure 3.1 Parallel system.

```
num1=1; den1=[1 2];
num2=[1 3]; den2=[1 10];
[numh,denh]=parallel(num1,den1,num2,den2);
printsys(numh,denh,'s')
```

MATLAB will respond with the resulting transfer function $H(s)$.

num/den =

$$\frac{s^2 + 6s + 16}{s^2 + 12s + 20}$$

3.5 Building Systems

The examples in the previous sections (with the exception of the last example) consist of systems described by a single transfer function or state-space model. Typically, control engineers will work with more complicated systems that involve many subsystems that are interconnected in various ways such as by feedback or in parallel or in series. Because control system analysis tools are most easily applied to a single transfer function or state-space model, MATLAB provides commands for reducing complex interconnected subsystems into a single description. In effect, MATLAB allows you to build a system from its pieces. The commands used for creating a single model out of its components include **cloop**, **feedback**, **series**, **parallel**, **blkbuild**, and **connect**.

The **cloop** command provides a means of generating a closed-loop (unity feedback) system description given an open-loop system. For example, consider the system shown in Figure 3.2.

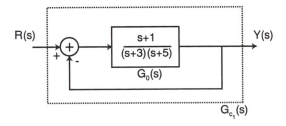

Figure 3.2 System with unity feedback.

File *closeunit.m* reduces the system to a single transfer function
$G_{c_1}(s) = G_0(s)/(1 + G_0(s))$.

closeunit.m
```
% closeunit
% Illustration of the use of cloop to create feedback
% systems.
ngo=[1 1];
dgo=conv([1 3],[1 5]);
[ngc1, dgc1]=cloop(ngo,dgo);
printsys(ngc1,dgc1,'s');
```

Entering **closeunit** at the MATLAB prompt produces $G_{c_1}(s)$ as

num/den =

$$\frac{s+1}{s\hat{}\,2 + 9s+16}$$

To close the loop for any kind of nonunity feedback, use the **feedback**
command. For example, the system of Figure 3.3 can be reduced using
the file *closefeed.m* to produce $G_{c_2}(s) = G_0(s)/(1 + G_0(s)H(s))$.

closefeed.m
```
% closefeed
% Using feedback to create a non-unity feedback system.
ngo=[1 1];
dgo=conv([1 3],[1 5]);
nh=[1 6];
dh=[1 10];
[ngc2,dgc2]=feedback(ngo,dgo,nh,dh);
printsys(ngc2,dgc2,'s')
```

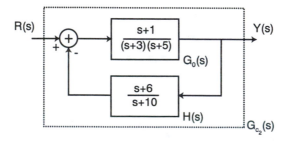

Figure 3.3 System with nonunity feedback.

Entering **closefeed** yields $G_{c_2}(s)$ as

num/den =

$$\frac{s\hat{}2 + 11\ s + 10}{s\hat{}3 + 19\ s\hat{}2 + 102\ s + 156}$$

For both **cloop** and **feedback,** negative feedback is the default. To specify positive feedback, use 1 as an additional, final argument to **cloop** or **feedback.** For example, to apply positive instead of negative unity feedback to the system of Figure 3.2 replace the third line of *closeunit.m* with **[ngc1,dgc1]=cloop(ngo,dgo,1);.**

The commands **cloop** and **feedback** can be used analogously to add feedback to state-space models. However, you must be consistent in the type of system description you use. That is, the input and output arguments to a single command must be either all transfer function descriptions or all state-space models.

In addition to **cloop** and **feedback,** the MATLAB commands **series** and **parallel** provide useful system reduction capabilities as described in Section 3.4. Figure 3.4 illustrates a system similar to the system of Figure 3.3 except that $H(s)$ is in series with $G_0(s)$ instead of in the feedback.

To reduce the system of Figure 3.4 to a single transfer function $G_{c_3}(s)$ we use both the **series** and **cloop** commands given in file *closeseries.m.*

closeseries.m
```
% closeseries
% System reduction using series and cloop.
ngo=[1 1];
dgo=conv([1 3], [1 5]);
```

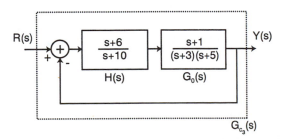

Figure 3.4 Feedback system with *H(s)* in series with *G₀(s).*

```
nh=[1 6];
dh=[1 10];
[nseries,dseries]=series(nh,dh,ngo,dgo);
[ngc3, dgc3]=cloop(nseries,dseries);
printsys(ngc3,dgc3,'s')
```

Using the four commands **cloop**, **feedback**, **series**, and **parallel**, you will be able to reduce a complicated system of interconnected subsystems into a single transfer function or state-space model. However, for very complicated systems, using these commands can become time-consuming and even confusing. As a result, MATLAB provides special commands to set up very complicated systems.

Setting up a very complicated system entails several steps with several MATLAB commands, which we illustrate by example. The system we will reduce to a single transfer function is shown in Figure 3.5. Suppose that $G_0(s) = 1, G_1(s) = 1/(s+1), G_2(s) = 1/(s+2), G_3(s) = 1/(s+3), H_1(s) = 4, H_2(s) = 8$, and $H_3(s) = 12$.

As you might imagine, using **cloop, feedback, series**, and **parallel** to reduce this system could get rather tedious. Instead we use the **blkbuild** and **connect** commands as shown in file *buildsystem.m*.

To begin with, we assign a number to each subsystem as shown in Figure 3.5. The first seven lines of *buildsystem.m* correspond to the descriptions of each of the seven subsystems according to the numbering system we selected. As we choose to describe each subsystem with a transfer function, we are required to name the corresponding numerators and denominators **n1, d1, n2, d2**, etc. Alternatively, if we had provided state-space models for the subsystems we would have had to name them **a1, b1, c1, d1, a2, b2, c2, d2**, etc.

buildsystem.m
```
% buildsystem
% Using blkbuild to construct a complicated system.
n1=1; d1=1;
n2=1; d2=[1 1];
n3=1; d3=[1 2];
n4=1; d4=[1 3];
n5=4; d5=1;
n6=8; d6=1;
n7=12; d7=1;
nblocks=7;
```

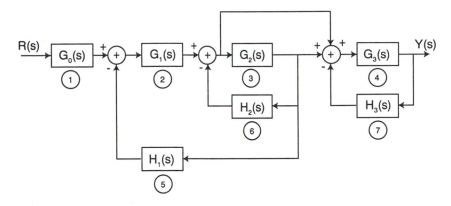

Figure 3.5 Block diagram of a complicated system.

```
blkbuild    %Note that this is an M-file, not a function.
q =    [2 1 -5 0  0
        3 2 -6 0  0
        4 2 -6 3 -7
        5 3  0 0  0
        6 3  0 0  0
        7 4  0 0  0];
inpt=1;
output=4;
[aa,bb,cc,dd]=connect(a,b,c,d,q,inpt,output);
[num,den]=ss2tf(aa,bb,cc,dd);
printsys(num,den,'s')
```

Once we have specified the seven subsystems, we assign the variable **nblocks** to be equal to the number of subsystems. The subsequent command **blkbuild** (note that this is an M-file, not a function) uses the variable **nblocks** to begin building the system. Specifically, **blkbuild** converts all the transfer function descriptions to state-space models using **tf2ss** and assembles them into one large block diagonal state-space model called **a,b,c,d** by using the **append** command repeatedly.

The next step is to create the matrix **q** that identifies the subsystem interconnections. Each row of the matrix **q** corresponds to a different subsystem. The first element in the row is the current subsystem number. The remaining numbers specify which blocks have output connected to the input of the current subsystem. For example, the first row of matrix **q** in *buildsystem.m* corresponds to subsystem 2 ($G_1(s)$), so the first element in this row is 2. Because the output of subsystem 1 ($G_0(s)$) and the negative output of subsystem 5 ($H_1(s)$) are input to subsystem 2, the next two elements in the row are 1 and −5. Two extra zeros at the end of the row are necessary to ensure that **q** is a rectangular matrix.

After defining **q**, we specify that subsystem 1 ($G_0(s)$) receives the system input and subsystem 4 ($G_3(s)$) produces the system output by defining the variables **inpt** and **output**. The **connect** command as shown in *buildsystem.m* makes the connections and reduces the system to a single state-space model.

At this point in our computations it would be appropriate to use **modred** and **balreal** to eliminate unnecessary states from our state-space model. However, these commands are more appropriately discussed with the material of Chapter 7, so we omit the discussion here.

To complete our model building routine, we convert the state-space model of the complete system back to a transfer function description using **ss2tf**. Running the file *buildsystem.m* generates the system transfer function for the combined system:

state model [a,b,c,d] of the block diagram has 7 inputs and 7 outputs
num/den =

$$\frac{-7.105e{-}15\ s\char94 2 + 1\ s + 3}{s\char94 3 + 26\ s\char94 2 + 179\ s + 210}$$

The first term in the numerator should actually be zero. This small error can be attributed to the numerical instability of the **tf2ss** and **ss2tf** commands. You should be aware of such potential errors when you build systems with transfer function descriptions. Using state-space models throughout is more reliable. In any case, it is generally a good idea to check that the final system description makes sense. It is particularly important to check the condition number of any matrices involved in the computation. It is also desirable to compare pole and zero locations (discussed in Chapter 4) for both transfer function and state-space descriptions. They should be the same.

3.6 Building Systems Using SIMULINK

SIMULINK is a MATLAB tool that provides a simple and easy-to-learn means of constructing systems from building blocks, driving them with a variety of inputs, and observing the results. As the name SIMULINK suggests, it is primarily a tool for simulating systems. It is particularly useful for studying the effects of nonlinearities on the behavior of systems.

The SIMULINK user interface is the standard Macintosh, PC, or Unix-based interface with multiple windows and pull-down menus. Most commands are available on pull-down menus. Many systems to be simulated can be constructed from simpler systems already available in SIMULINK. The building blocks are represented by icons. These icons (simple systems or directories containing a set of such systems) are opened by double-clicking on the icon and moved by depressing the mouse button and dragging. Anyone who has used a Macintosh or a similar user interface already knows the basics of using SIMULINK. We present an introduction by means of an example and some suggestions on how to use SIMULINK in the following.

To start SIMULINK it is necessary to start MATLAB. Once MATLAB is running, simply enter the command **simulink** and wait. Shortly, a box with seven icons, five pull-down menu heads, and the word "simulink" in its header bar will appear on the screen. Use the mouse to move the mouse arrow to the **file** menu head, pull down the menu, and click (always use the left mouse button) on **new**. This will open another window. Note that this window has the word "untitled" in its header bar. Use the mouse to drag it to a convenient place.

We now construct a system in this window. Go back to the first SIMULINK window and double-click on the **linear** icon. This will open a window with (currently) ten icons in it and the title "linear." Click and drag the **transfer Fcn** icon, into the "untitled" window. It is important to click near the center of the icon because clicking elsewhere has a different effect. Position it as shown in Figure 3.6.

Next, click and drag the **gain** icon to the position shown in Figure 3.6. Repeat this process to place the **sum** and **integrator** icons in the "untitled" window also. This is a good time to pull down the **file** menu in the "untitled" window and click on **save as**. A dialog box will appear. Click in the rectangular box below the word selection and type **simplsim**. Then, click on the **done** icon or strike the return key to close the dialog box. Notice that the "untitled" window is now the "simplsim" window. Close the "linear" window.

Double-click on the **sources** icon in the "simulink" window to open it. Click and drag the **step input** into the "simplsim" window. Close the "sources" window. Double-click on the **sinks** icon and drag the **graph** icon from the "sinks" window into the "simplsim" window. Pull down the

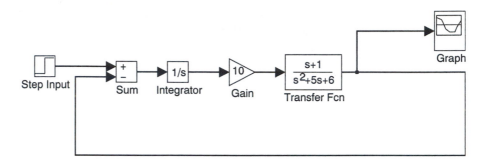

Figure 3.6 SIMULINK block diagram as displayed in the "simplsim" window.

file menu in the "simplsim" window and click on **save**.

To connect the icons in the "simplsim" window, depress the mouse button while the mouse arrow is on the output symbol of an icon, drag to the input symbol of the icon to which you want to connect, and release. To make a corner in a connection simply release the mouse button, click again on the end of the current line, and drag. Put all the connections into "simplsim" in this way and save the result using the **file** menu and **save** as before.

If you don't like some aspect of the picture, you can change it in a variety of ways. You can move any of the icons by clicking on its center and dragging. You can move any of the lines by clicking on one of its corners and dragging. You can change the size and shape of any of the icons by clicking and dragging one of its corners. You can remove any line or icon by clicking on it to select it, pulling down the **edit** menu, moving to **cut**, and releasing the mouse button. Once you are happy with "simplsim," save it.

Double-click on the center of the **Transfer Fcn** icon in "simplsim." A dialog box will appear. Click and wipe in the rectangle below **numerator** to select everything in that rectangle. Type **[1 1]** and then click and wipe in the **denominator** rectangle. Type **[1 5 6]** and then click in the **done** icon. You may find that you have to drag a corner of the **Transfer Fcn** icon to make it large enough to display the whole transfer function as done in Figure 3.6. Double-click on the **Gain** icon and set the gain to 10.

Double-click on the **Sum** icon and set the signs to + −. In this example we leave the **Step Input** at its default values. Double-click on the **Graph** icon. Set the time range to 3, y-min to 0, and y-max to 2. Save "simplsim."

Pull down the **Simulation** menu and release the mouse on **Parameters**. Set the start time to 0.0 and the stop time to 3. Leave the other parameters at their default values. Save "simplsim." You should now have exactly the same system as shown in Figure 3.6.

To run a simulation of the system you created, pull down the **Simulation** menu again and release on **Start**. SIMULINK will create a graph window and display in it the response of this system to a unit step at $t = 1$.

Additional SIMULINK examples can be found in subsequent chapters. We encourage you to create your own.

3.7 Exercises

1. Consider a mass, spring, and damper system with mass $M = 2$, spring constant $K = 3$, and damping coefficient $\mu = 0.8$. Find the second-order transfer function $G(s) = X(s)/F(s)$ where $X(s)$ is position of the mass and $F(s)$ is the force applied to the mass. What are the natural frequency and normalized damping ratio of the system? Find a state-space model of the system.

2. (a) Suppose a system described by the transfer function $H(s)$ is connected in parallel with the system in problem 1 described by $G(s)$. Write a script file that produces the transfer function of the system given that $H(s)$ is a random, stable third-order system.

    (b) Repeat part (a) using state-space models of each of the systems in parallel.

3. (a) Consider the system of Figure 3.3. Suppose, however, that $H(s)$ provides positive feedback instead of negative feedback. Denote the transfer function of the new combined system by $G'_{c_2}(s)$. Find $G'_{c_2}(s)$.

    (b) What are the roots of the denominator polynomial of $G'_{c_2}(s)$? How do they compare to the roots of the denominator polynomial of $G_{c_2}(s)$?

4. Consider a plant controlled by negative feedback where $G_{\text{plant}}(s)$ is the transfer function description of the open-loop plant and $G_{\text{ctrl}}(s)$

is the transfer function description of the controller that is placed in the feedback loop. Let $G_{\text{plant}}(s) = H(s)G_{c_1}(s)$ where $H(s)$ is as shown in Figure 3.1 and $G_{c_1}(s)$ is as shown in Figure 3.2. Let $G_{\text{ctrl}}(s) = G_{c_3}(s)$ where $G_{c_3}(s)$ is as shown in Figure 3.4.

(a) Find the transfer function of the closed-loop system. Use the commands **cloop, feedback, series**, and **parallel**.

(b) Find the transfer function of the closed-loop system using the commands **blkbuild** and **connect**.

5. (a) Use SIMULINK to create a closed-loop system similar to that in Figure 3.6 with open-loop transfer function $G(s) = (s+1)/(s+2)(s+3)$ in place of $(s+1)/(s^2 + 5s + 6)$ and a saturation nonlinearity in place of the integrator.

(b) Simulate the step response of this system and display the results in **Graph**.

(c) Add another **Graph** block after the saturation to see what the signal there looks like.

4 POLES, ZEROS, AND TIME RESPONSE

This chapter describes the MATLAB tools for studying the relation between the poles and zeros of a linear time-invariant system and its time-domain responses. Because control system specifications are very often given in terms of the response to a unit step, we emphasize the step response. Because of the theoretical importance of the impulse response, it also receives considerable attention.

4.1 Computing Poles and Zeros

Determining the poles and zeros of a given transfer function $G(s)$ by hand calculation can be an unpleasant and time-consuming job. It requires factoring two polynomials. The usual academic examples, involving second- or third-degree polynomials with integer factors, hide the difficulty. So does MATLAB, by performing the calculation in response to a single command, **tf2zp**. We illustrate the ease with which MATLAB determines poles and zeros with the script file *findingpz.m*, which computes the poles and zeros of the transfer function

$$G(s) = \frac{(3s^4 + 2s^3 + 5s^2 + 4s + 6)}{(s^5 + 3s^4 + 4s^3 + 2s^2 + 7s + 2)}$$

If you doubt that this is a difficult calculation to do without the computer you can try this example by hand. In fact, this example cannot be solved by direct application of formulas for the roots of polynomials. The roots of polynomials of degree five or higher can be found only numerically.

findingpz.m

```
% findingpz
% This creates a transfer function and then finds and
% displays its poles, zeros, & gain.
num=[3 2 5 4 6];        % create numerator polynomial
den=[1 3 4 2 7 2];      % create denominator polynomial
[z,p,k]=tf2zp(num,den) % find & display poles & zeros
```

The MATLAB response to running file *findingpz.m* is given below.

```
z =
      0.4019 + 1.1965i
      0.4019 - 1.1965i
     -0.7352 + 0.8455i
     -0.7352 - 0.8455i
p =
    -1.7680 + 1.2673i
    -1.7680 - 1.2673i
     0.4176 + 1.1130i
     0.4176 - 1.1130i
    -0.2991
k =
     3
```

The variables **num** and **den** describe the numerator and denominator polynomials of $G(s)$, respectively. MATLAB returns the zeros, denoted by **z**, the poles, denoted by **p**, and the gain, denoted by **k**, of $G(s)$ in response to the command **tf2zp** as called in the file. Despite the fact that the coefficients of the numerator and denominator polynomials are integers and positive, the poles and zeros are not integers. The system is also unstable (has poles in the right half-plane) and nonminimum phase (has zeros in the right half-plane).

MATLAB will also compute the poles and zeros from a state-space description of the system. We illustrate the use of the command to do this, **ss2zp**, in the file *findingsszp.m*.

findingsszp.m

```
% findingsszp
% This creates a state-space system and then finds and
% displays its poles, zeros, & gain.
vec=[1 2 5 3 6 1]; % Create a fifth order polynomial
A=compan(vec);     % Create the companion matrix
```

```
c=[1 0 0 0 0];          % Create the output vector
b=[0;0;0;0;1];          % Create the input vector
[z,p,k]=ss2zp(A,b,c,0)  % Find & display poles, zeros,
                        % & gain
```

The system of interest is described by the state-space model (**A**, **b**, **c**, **0**). Notice the use of **compan** to create the companion matrix **A** from its characteristic polynomial **vec**. The matrix **A** has dimension one less than the vector that describes its characteristic polynomial. Running the file produces the response

```
z =
    1.0e-05*
    -0.1455 + 0.2520i
    -0.1455 - 0.2520i
     0.2910
p =
    -1.0433 + 1.6215i
    -1.0433 - 1.6215i
     0.1323 + 1.2217i
     0.1323 - 1.2217i
    -0.1781
k =
    -1.0000
```

There is an odd response immediately after **z=**. The symbols **1.0e–05*** indicate that all the responses for the zeros are to be multiplied by 0.00001 in order to obtain their correct values. That is, this system has three zeros, and they are all very close to zero. To see these zeros as well as the poles of the system on a plot of the complex plane, simply enter the command **pzmap(p,z)** in the command window right after running file *findingsszp.m*. The result is the pole-zero plot of Figure 4.1. Notice that five poles appear but only one zero and it is at zero. This is a minor flaw in the graph. As you know, there are three zeros, but they are virtually on top of each other. Evidently, the multiplicity of poles and zeros is not indicated by **pzmap**.

You should be wondering whether the computed poles and zeros are correct. It is easy to verify that the n-dimensional system with companion form state-space representation (as produced by *findingsszp.m*)

$$\dot{x}_1(t) = -a_1 x_1(t) - a_2 x_2(t) \cdots -a_n x_n(t)$$
$$\dot{x}_2(t) = x_1(t)$$

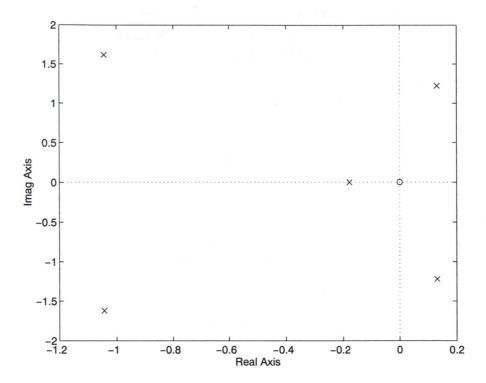

Figure 4.1   Pole-zero map for the system created by file *findingsszp.m*.

$$\dot{x}_3(t) = x_2(t) \tag{4.1}$$

$$\vdots$$

$$\dot{x}_{n-1}(t) = x_n(t)$$
$$\dot{x}_n(t) = x_{n-1}(t) + u(t)$$

$$y(t) = x_1(t) \tag{4.2}$$

does indeed have $(n - 2)$ zeros at the origin and must have gain $-a_n$.

Taking the Laplace transform of Eqs. (4.1) and (4.2) yields

$$sX_1(s) = -a_1 X_1(s) - a_2 X_2(s) \cdots - a_n X_n(s)$$
$$sX_2(s) = X_1(s)$$
$$sX_3(s) = X_2(s) = X_1(s)/s \tag{4.3}$$

$$\vdots$$

$$sX_{n-1}(s) = X_1(s)/s^{n-3}$$
$$sX_n(s) = X_1(s)/s^{n-2} + U(s)$$

$$Y(s) = X_1(s) \tag{4.4}$$

Eliminating all the $X_i(s)$ from the first of Eqs. (4.3) then gives

$$sY(s) = -a_1 Y(s) - a_2 \frac{Y(s)}{s} \cdots - a_{n-1} \frac{Y(s)}{s^{n-2}} - a_n \frac{Y(s)}{s^{n-1}} - \frac{a_n U(s)}{s}$$

or

$$(s^n + a_1 s^{n-1} + \cdots + a_n) Y(s) = -a_n s^{n-2} U(s) \qquad (4.5)$$

Equation (4.5) shows that the gain is $-a_n$ and there are $n - 2$ zeros at the origin. Thus, we confirm for our example that $k = -1$ and there are three zeros at the origin.

The command **pzmap** could also have been used to do almost the same calculations that were done with **ss2zp** and **tf2zp** in our first two examples. The commands **[p,z]=pzmap(A,b,c,0)** and **[p,z]= pzmap(num,den)** take, respectively, the state-space and transfer function descriptions and return the poles and zeros. Notice that they do not return either the gain (which is why it is probably better to use **ss2zp** and **tf2zp**) or a pole-zero plot. There is a separate command, **dcgain**, that will return the dc gain. The dc gain is slightly different from the gain. Given $G(s) = k(s+2)/(s+1)(s+4)$ the gain is k but the dc gain is $k/2$. This is because the gain is the coefficient of the highest power of s in the numerator when the denominator is monic (the coefficient of the highest power of s is 1). The dc gain is the output amplitude when the input is a unity dc signal. If you leave off the left-hand-side arguments, such as **pzmap(num,den)**, then MATLAB produces the pole-zero plot as shown in Figure 4.1.

4.2 Creating Transfer Functions from Poles and Zeros

It is easy to go to a transfer function description from a pole-zero description of a linear time-invariant system z, p, k where z is a vector of zeros, p is a vector of poles, and k is the scalar gain. This is done by means of the command **zp2tf** and illustrated for a system with four zeros, five poles, and a gain of 2 in the file *zeropole2trans.m*. **zp2tf** was designed to work with a matrix containing the zeros of a single-input, multiple-output (SIMO) system. For a SISO system, the zero matrix z must be a column vector. The vector of poles can be a row or column vector. We also illustrate the use of **printsys** to display the resulting transfer function explicitly.

zeropole2trans.m
```
% zeropole2trans
```

```
% This creates a transfer function from its poles,
% zeros, & gain.
% It displays the transfer explicitly.
z=[0; 0; 0;-1];
p=[-1+i  -1-i  -2  -3.15+2.63*i  -3.15-2.63*i];
k=2;
[num,den]=zp2tf(z,p,k);
printsys(num,den,'s')
```

Running the file produces the output

num/den =

$$\frac{2\ s\char`^4 + 2\ s\char`^3}{s\char`^5 + 10.3\ s\char`^4 + 48.04\ s\char`^3 + 109.2\ s\char`^2 + 126.2\ s + 67.36}$$

It is also possible to use MATLAB to convert a list of poles, zeros, and gain to a state-space description. This is accomplished by means of the command **zp2ss** as illustrated in file *examplezp2ss.m* for a system with one zero at -1, three poles (one each at -2, $-1 + i$, and $-1 - i$), and a gain of 2.

examplezp2ss.m

```
% examplezp2ss
% This creates a state space system from its poles
% & zeros.
[a,b,c,d]=zp2ss([-1],[-2  -1+i  -1-i],2)
pause                   % stops the file from running
poles=eig(a)            % computes poles
pause
zros=tzero(a,b,c,d)     % computes zeros - zeros is a
                        % MATLAB function
pause
k=dcgain(a,b,c,d)       % computes dc gain
```

The file introduces several other commands. The command **eig** computes the eigenvalues of its matrix argument. As you can verify by running the file, the eigenvalues of **a** are the poles of the system; applying the command **tzero** computes the zeros; and **dcgain** computes the dc gain. We have used the **pause** command to give you a chance to study the results. When the script file reaches **pause**, it stops running. Striking any key will cause the script file to resume operating. Note that the command **tzero** will compute several different kinds of zeros for a multiple-input, multiple-output system. SISO systems have only one kind of zero so we will not discuss the more complex uses of this command.

As we repeatedly emphasize, the MATLAB routines are generally reliable. However, you do have to be cautious about the numerical accuracy of the computation of poles and zeros. The roots (zeros or poles) of high-order systems and roots with large multiplicity are particularly difficult to compute. File *backandforth10.m* was created to demonstrate the problem. The system has no zeros, 10 poles at $s = -1$, and a gain of 1.

backandforth10.m

```
% backandforth10
% Example of the possible problems in
% computing poles and zeros.
z=[];
p=[-1  -1  -1  -1  -1  -1  -1  -1  -1  -1];
k=1;
[num,den]=zp2tf(z',p',k)
[z1,p1,k1]=tf2zp(num,den)
```

Notice that the vector of zeros **z** is specified by an empty vector **[ ]** since there are no zeros in this example. The command **zp2tf** is used to convert the zero-pole-gain system description **z**, **p**, **k** into a transfer function described by numerator polynomial **num** and denominator polynomial **den**. The command **tf2zp** is then used to convert the transfer function back into a zero-pole-gain description **z1**, **p1**, **k1**, which can be compared to the original description **z**, **p**, **k**. Running the file produces the following result.

```
num =
     0   0   0   0   0   0   0   0   0   0   1
den =
     1  10  45 120 210 252 210 120  45  10   1
z1 =
     []
p1 =
    -1.0474
    -1.0376 - 0.0284i
    -1.0376 + 0.0284i
    -1.0130 - 0.0445i
    -1.0130 + 0.0445i
    -0.9846 - 0.0428i
    -0.9846 + 0.0428i
    -0.9633 - 0.0256i
```

```
      -0.9633 + 0.0256i
      -0.9555
k1 =
      1
```

The numerical values of **p1** are machine dependent, so you may get slightly different values if you run *backandforth10.m* yourself.

Notice that the denominator polynomial **den** looks fairly safe. There is just slightly more than a factor of 100 difference between the largest and smallest coefficients. However, several of the computed poles **p1** are in error by about 5%.

4.3 Computing and Plotting Time Response from Poles and Zeros

One of the strengths of MATLAB is the ease with which it creates graphs of the time-domain response of a linear time-invariant system. In this section we introduce the commands **step, impulse, initial, lsim**, and **damp**. If the poles, zeros and gain are given, it takes two lines of code to produce either the impulse response (using **impulse**) or the step response (using **step**). This is illustrated in file *simple.m* for a system with no zeros, poles at $s = -1 \pm 3i$, and a gain of 3.

simple.m
```
% simple
% A simple example of displaying the step response.
[num,den]=zp2tf([],[-1+3*i   -1-3*i],3);
step(num,den)
```

The resulting graph is shown in Figure 4.2.

Control systems are often required to satisfy specifications on several aspects of their step response. Typical examples are:

1. Time to peak = the time for the step response to reach its first (and largest) peak.

2. Percent overshoot = $100 \left(\dfrac{\text{peak value} - \text{final value}}{\text{final value}} \right)$

3. Rise time = time at which step response first reaches 90% of its final value minus time at which it first reaches 10% of its final value.

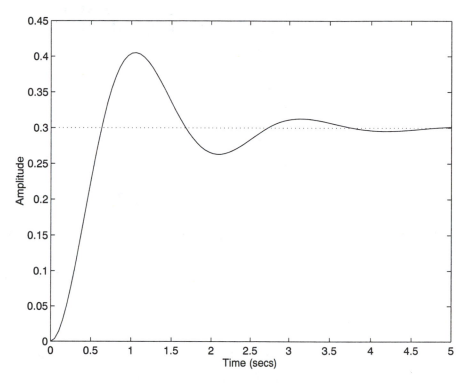

Figure 4.2   Step response of $G(s) = 3/(s + 1 - 3i)(s + 1 + 3i)$.

4. Settling time = last time the step response is farther than ±2% from its
   final value.

Note that people sometimes use other percentages to define rise time and
settling time.

We have written a script file, *stepanalysis.m*, to compute all of the fore-
going parameters of the step response shown in Figure 4.2.   Note the
computation of **finalvalue** on line 2 of the file. This is a direct implementation of the formula

$$\text{Final value} \ = \lim_{s \to 0} sG(s)\frac{1}{s} = G(0).$$

MATLAB will tell you when there are problems with this formula. For example, you will be warned about **NaN** or **0/0** if $G(s)$ has a pole at zero.

stepanalysis.m

```
% stepanalysis
% This creates a simple transfer function and
% performs an analysis of its step response.
```

```
[num,den]=zp2tf([],[-1+3*i   -1-3*i],3);
finalvalue=polyval(num,0)/polyval(den,0)
[y,x,t]=step(num,den);
[Y,k]=max(y);
timetopeak=t(k)
percentovershoot=100*(Y-finalvalue)/finalvalue
% compute rise time
n=1;
    while y(n)<0.1*finalvalue, n=n+1; end
m=1;
    while y(m)<0.9*finalvalue, m=m+1; end
risetime=t(m)-t(n)
% compute settling time
l=length(t);
    while (y(l)>0.98*finalvalue)&(y(l)<1.02*finalvalue)
    l=l-1;
    end
settlingtime=t(l)
```

Line 3 of *stepanalysis.m* computes the step response using **step** with output arguments and stores it in the vector **y**. The time vector corresponding to **y** is stored in the vector **t**. [The vector **x** represents the (MATLAB-selected) state as a function of time. However, we are not interested in the state in this example.] Line 4 uses the command **max** to find the peak value of **y** and its index. These are stored as **Y** and **k**. The next two lines compute and display the time to peak and the percent overshoot. Note that these two lines of code assume that you have already seen Figure 4.2 and know that the first peak in the step response is also the largest peak. In any case, time to peak is not a good performance measure for systems whose step response requires several oscillations to reach its highest peak. See Figures 4.6 and 4.7 for examples of such systems.

As the comment in the file indicates, the next several lines of code compute the rise time. Note the use of the **while** loops to compute one less than the indices of the first values of **y** greater than **0.1*finalvalue** and **0.9*finalvalue**.

The settling time computation again uses the fact that we have seen Figure 4.2 and know that there is a settling time. We work backward from the index of the last time in the plot [**l** = **length(t)**] to the settling time.

The result of running file *stepanalysis.m* is

```
finalvalue =
     0.3000
timetopeak =
     1.0606
percentovershoot =
     35.0607
risetime =
     0.4545
settlingtime =
     3.5354
```

It is also possible to use MATLAB to visualize the effect of changes in the poles and zeros of a transfer function on the system response. We created the script file *meshplot.m* to demonstrate the effect of the change in pole location on the step response of a second-order system. The file creates a mesh plot of 12 step responses for a sequence of systems with poles located at $s = -n/4 \pm 3i$ as n varies from 1 to 12. A very similar plot appears on the cover of the old MATLAB Control System Toolbox manual. The first two lines of code in the file create a vector, t, of time ranging from 0 to 5 and the number of curves to be plotted (we have chosen 12).

meshplot.m
```
% meshplot
% This creates a mesh plot showing the effect of
% increasing
% the real part of a pair of complex conjugate poles.
clf
t=[0:0.05:5];
numberofcurves=12;
y=zeros(length(t),numberofcurves);
n=1;
while n<=numberofcurves,
     [num,den]=zp2tf([],[-n/4+3*i    -n/4-3*i],
     (n/4)^2+9);
     [y(1:length(t),n),x,tdumb]=step(num,den,t);
     n=n+1;
end
mesh(t,1:12,y')
title('Mesh Plot Showing Step Response for Twelve Pole
     Locations')
```

Mesh Plot Showing Step Response for Twelve Pole Locations

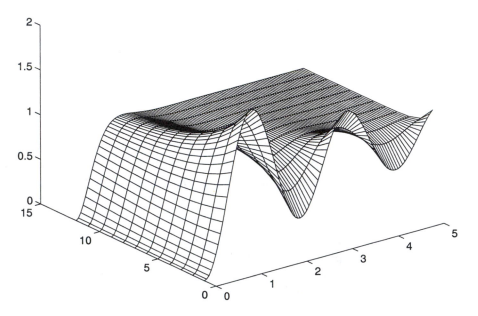

Figure 4.3   Mesh plot created by *meshplot.m*.

The next line creates a matrix, **y**, with all elements equal to zero. If you were to delete this line of code and run the file you would still get the same result we do. That is, you would get Figure 4.3. This version of *meshplot.m* is the fifth one we tried. The earlier versions did not produce as nice a figure. In particular, mesh plots seem to look better if there are fewer points. An earlier version contained a much larger matrix **y**. As a result, when we ran the current version without the fourth line of code, the matrix **y** was not redimensioned, i.e., MATLAB replaced the first 12 columns of the matrix but did not change the remaining columns. The result was a very bizarre mesh plot. We decided to include the line of code that reset **y** for two reasons. We wanted to remind you that MATLAB retains the variables that it computes until you tell it to discard them or you quit MATLAB. Additionally, if you do enough work with MATLAB you will run into a problem just as we did and will need to know how to fix it. This demonstrates one way to deal with the problem. Alternatively, you could enter the command **clear y** to remove the previously defined **y** or just **clear** to remove all the variables from the workspace.

Lines 5 through 10 of *meshplot.m* compute the step responses for complex conjugate pairs of poles ranging from -0.25±3i to -3±3i. The imaginary part of the poles is held fixed at ±3i. The real part is varied in increments of 0.25. **zp2tf** and **step** are used to compute the step response. Notice that the last argument of **zp2tf**, which corresponds to the gain, ensures that all the step responses have final value equal to one. That is, we compute the gain needed to make the dc gain 1. The resulting step responses are stored in **y**. The last three lines in the file create and title the mesh plot. Note that the title must appear on a single line. MATLAB will respond with an error message if you duplicate the last two lines of *meshplot.m.*

Figure 4.3 shows the change in the step responses. When the real part of the poles is –0.25 the step response is very oscillatory (front of plot). As the poles are moved to the left, the step response gradually becomes smoother (back of plot). Notice also that the rise time decreases as the poles move to the left. In other words, the oscillatory responses reach the final value faster. The curve farthest to the back shows what is generally agreed to be a good compromise between speed and smoothness. This curve corresponds to poles at –3±3i which correspond to a damping ratio of 0.707.

MATLAB also facilitates the computation of both damping ratio and natural frequency. We illustrate this with script file *damping.m* for a system with no zero, poles at $-1 \pm ni$, and unit dc gain for n ranging from 1 to 4. The command **damp** returns the roots of the denominator polynomial of the transfer function (equivalent to pole locations) and damping ratio and the natural frequencies of all the poles in a given polynomial. **damp** refers to the poles as eigenvalues because **damp** can also be used to find the poles of a state-space model by finding the eigenvalues of *A*.

damping.m

```
% damping
% This illustrates the use of damp.
number_of_tests=4;
n=1;
while n<=number_of_tests,
     [num,den]=zp2tf([],[-1+i*n   -1-i*n],(n^2+1));
     damp(den)
     n=n+1;
end
```

The result of running *damping.m* is shown next. Notice how the damping ratio and natural frequency change as we move the poles.

Eigenvalue	Damping	Freq. (rad/sec)
-1.0000 + 1.0000i	0.7071	1.4142
-1.0000 - 1.0000i	0.7071	1.4142
Eigenvalue	Damping	Freq. (rad/sec)
-1.0000 + 2.0000i	0.4472	2.2361
-1.0000 - 2.0000i	0.4472	2.2361
Eigenvalue	Damping	Freq. (rad/sec)
-1.0000 + 3.0000i	0.3162	3.1623
-1.0000 - 3.0000i	0.3162	3.1623
Eigenvalue	Damping	Freq. (rad/sec)
-1.0000 + 4.0000i	0.2425	4.1231
-1.0000 - 4.0000i	0.2425	4.1231

Most textbooks emphasize first- and second-order systems for the good reason that the relation between pole location and time response can be stated precisely. Furthermore, many real systems can be very well approximated by systems with one or two poles. We created file *approx.m* to demonstrate how well third-order systems can be approximated by second-order systems. In this example we consider a system with no zeros, poles at $s = -1 \pm 3i$ and at $s = -1 - n$, and unit dc gain. We vary n from 1 to 12. We plot the step responses of these 12 systems side by side on a mesh plot. We also plot, at the back, the step response of the second-order system with poles at $s = -1 \pm 3i$. We can then use this plot to see how closely the second-order system approximates the third-order systems.

approx.m

```
% approx
% This creates a mesh plot that shows how good the
% second-order approximation can be.
t=[0:0.05:5];
numberoftests=12;
y=zeros(length(t),numberoftests);
n=1;
while n<=numberoftests,
   [num,den]=zp2tf([], [-1+3*i  -1-3*i  -1-n],10*(n+1));
   [y(1:length(t),n),x,tdumb]=step(num,den,t);
   n=n+1;
end
[numex,denex]=zp2tf([],[-1+3*i  -1-3*i],10);
```

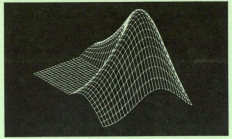

MATLAB® Technical Computing Environment

MATLAB is used in **Using MATLAB to Analyze and Design Control Systems, 2/e** by Naomi Ehrich Leonard and William S. Levine (Benjamin/Cummings, 1995) for problem-solving. The MATLAB M-files in the *Analyze and Design Control Systems Toolbox* have been created by the authors to illustrate the concepts presented in the text.

- **MATLAB Application Toolboxes** add functions for symbolic math, signal processing, control design, neural networks, and other areas.

- **The Student Edition** is a limited-matrix-size version of MATLAB for use on students' own personal computers.

- **Educational discount plans** support classroom instruction and research.

- **Classroom Kits** provide cost-effective support for PC or Mac teaching labs.

- **MATLAB-based books** use MATLAB to illustrate basic and advanced material in a wide range of topics.

The MATH WORKS Inc.

BUSINESS REPLY MAIL

FIRST CLASS MAIL   PERMIT NO. 82   NATICK, MA

POSTAGE WILL BE PAID BY ADDRESSEE

The MathWorks, Inc.
24 Prime Park Way
Natick, MA 01760-9889

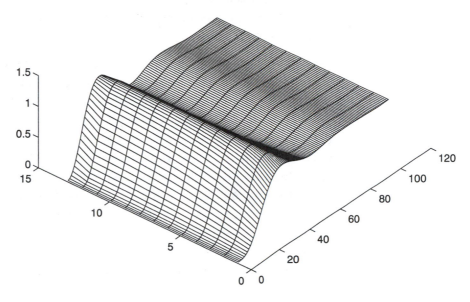

Figure 4.4 Mesh plot showing approximation by second-order systems.

```
[y(1:length(t),13),x,tdumb]=step(numex,denex,t);
clf
mesh(t,1:13,y'), view([-50 60])
```

As should be obvious, we created the file *approx.m* by editing *meshplot.m*. We fixed the complex conjugate pole pair at –1±3*i*; we added a pole at –(1 + *n*); and we changed the normalization so all the step responses would have one as their final value. We also added a 13th column to the matrix **y**. This column contains the exact second-order response. Finally, we added the **view** command. The **view** changes the viewpoint from which you see the plot. Figure 4.4 shows the mesh plot of step responses. The step response farthest to the front corresponds to the system with the third (real) pole closest to the pair of complex poles. Comparing Figures 4.3 and 4.4, you will note that the viewpoint in Figure 4.4 is both higher above the plot and more behind the *t* = 0 line. The default viewpoint, shown in Figure 4.3, is [–37.5 30]. Compare this to our choice.

Notice that only the lowest few values of *n* (corresponding to the curves at the front of the plot) produce noticeably different step responses from that for *n* = 12. The exact response, the one farthest to the back, is slightly different from the plot where *n* = 12. The difference is so small that you would not notice it if the curves were not side by side.

There is much more that could be learned about the relation between poles and zeros and step responses by drawing mesh plots showing the change in response as a function of changes in parameters. We leave this as a series of Exercises.

Because second-order systems play such an important role in control design, we demonstrate a plotting tool known as a phase-plane plot that can be very useful for providing insight into the behavior of such systems. File *phaseplane.m* produces the phase-plane plots shown in Figure 4.5, which illustrate the dynamics of a second-order system. The command **ord2** is used to provide a transfer function model of a second-order system having natural frequency $w_n = 10$ and damping ratio $\zeta = 0.1$. The phase plane for a second-order system corresponds to choosing state vector $x = (x_1, x_2)$ with $x_2 = dx_1/dt$. The command **step** is used to compute the step response of the state vector **x**. Then, the second component of the state vector **x(:,2)** is plotted against the first component **x(:,1)** for all time **t** using **plot**. A similar plot is created for the impulse response using **impulse**.

phaseplane.m
```
% phaseplane
% This creates a second-order system and displays phase
% plane plots of its step and impulse responses.
clf
[num,den]=ord2(10,.1);
[y,x,t]=step(num,den);
subplot(211)
plot(x(1:length(t),2),x(1:length(t),1))
xlabel('x1-axis'), ylabel('x2-axis'),
title('x2 vs. x1 for step input'),
[y1,x1,t1]=impulse(num,den);
subplot(212),
plot(x1(1:length(t),2),x1(1:length(t),1))
xlabel('x1-axis'), ylabel('x2-axis'),
title('x2 vs. x1 for impulse input'),
```

Plotting the states x_2 vs. x_1 shows the trajectory, or path, traversed through the state-space. Notice in the top plot that the response to a step input spirals to the point $x_2 = 0, x_1 = 0.01$ while the impulse response in the bottom plot spirals to (0,0). Notice also the different starting points, (0,0) for the step input and $x_2 = 1, x_1 = 0$ for the impulse input. Finally, note that the curves are always vertical when they pass through the $x_2=0$ line. This is because $x_2 = dx_1/dt$. When $x_2 = 0$, x_1 cannot be changing.

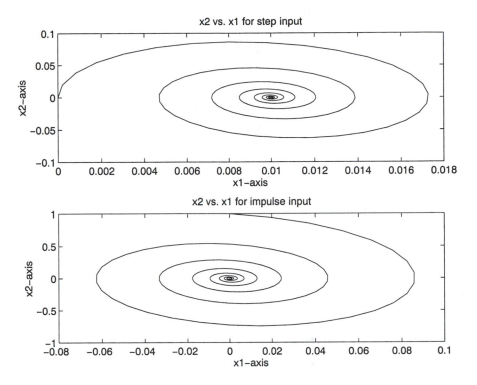

Figure 4.5 Plots of the phase-plane trajectories of $G(s) = 1/(s^2 + 2s + 100)$ in response to a unit step and a unit impulse.

The state-space trajectories for critically damped, overdamped, and unstable systems are very different from those shown in Figure 4.5. The Exercises include creation of state-space trajectories for such systems.

For some second-order systems, the phase-plane trajectory displays the flow of energy between kinetic and potential (or between electrical and magnetic) in a particularly graphic way. Suppose our system consists of a mass, spring, and dashpot as in *massspringdamp.m* in Chapter 3. If x_1 represents position and x_2 velocity, then the kinetic energy is zero when $x_2 = 0$ and the potential energy is zero when $x_1 = 0$. Every time the state-space trajectory spiral completes 360° of rotation you can see the amount by which the total energy of the system has decreased.

The following series of examples illustrate the use of MATLAB to determine impulse responses using **impulse**, initial condition responses by means of **initial**, and responses to more general input waveforms using **lsim**. We also use the examples to illustrate several aspects of system theory.

The file *stepvimp.m* is primarily intended to demonstrate the difference between the impulse response and the step response. The difference is particularly dramatic when the system has a pair of lightly damped poles as well as a single real pole with a relatively long time constant. Thus, we demonstrate the responses for a system with three poles at $s = -1$ and $s = -0.2 \pm 10i$. The system has no zeros and a gain of 100. The file *stepvimp.m* is given next and is followed by the resulting plots in Figure 4.6. Notice that the oscillation is much more pronounced in the impulse response than in the step response. The step input puts much less energy into the oscillatory component with the result that the response is much smoother. We will return to this point in the Exercises and in later chapters.

stepvimp.m

```
% stepvimp
% Comparison of step vs. impulse response for a
% third order system with 2 lightly damped poles.
clf
[num,den]=zp2tf([],[-1 -0.2+10*i -0.2-10*i],100);
subplot(211), step(num,den)
subplot(212), impulse(num,den)
```

The file *sonofstepvimp.m* illustrates the effect of a zero on the same system. The main difference between the system in file *sonofstepvimp.m* and the system in file *stepvimp.m* is the zero at -2 in the second file. The other differences between the files are commands to customize the plots. We wanted to force the time scale to be the same in both plots. Note that this is not true in Figure 4.6.

sonofstepvimp.m

```
% sonofstepvimp
% This shows the effect of adding a zero to the example
% in stepvimp.m.
clf
[num,den]=zp2tf([-2],[-1 -0.2+10*i -0.2-10*i],100);
[y,x,t]=step(num,den);
[yi,xi,t]=impulse(num,den,t);
subplot(211), plot(t,y)
title('step response'),xlabel('time'),ylabel('amplitude')
subplot(212), plot(t,yi)
title('impulse response'),xlabel('time'),
ylabel('amplitude')
```

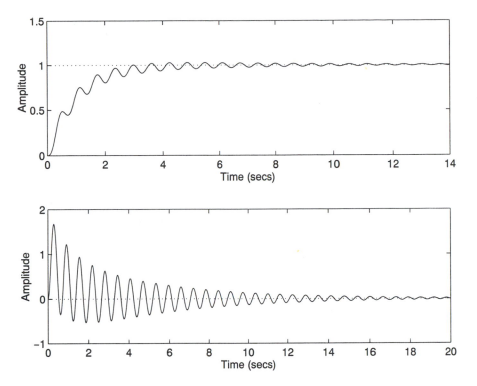

Figure 4.6 Step and impulse responses of $G(s) = 100/(s+1)(s+0.2-10i)$ $(s+0.2-10i)$.

We also wanted to demonstrate slightly better labeling. The plots that result from running *sonofstepvimp.m* are shown in Figure 4.7. Notice the change in steady-state value of the step response. It is now two instead of one. Also, notice that the zero has greatly enhanced the effect of the lightly damped pair of poles.

The next example also contains three poles and one zero but we have moved the complex conjugate pair of poles to both higher frequency and greater damping ratio. After plotting the step response in file *stepandic.m* we plot the initial condition response for two different initial conditions on the same graph using the command **initial**. The result is shown in Figure 4.8. Notice the two uses of the command **initial**, first without a left-hand argument and then with one. The command **initial** is used to calculate the initial condition response. Used without left-hand arguments, **initial** automatically generates a plot. We have also labeled the graphs and used **plot** to specify a different line type for the second initial condition response. Note, in Figure 4.8, that there is a large difference in the extent to which the two initial conditions excite the system.

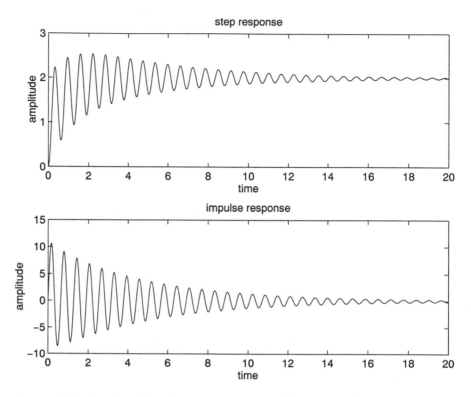

Figure 4.7   Step and impulse responses of $G(s)=100(s+2)/(s+1)(s+0.2-10i)$ $(s+0.2+10i)$.

stepandic.m

```
% stepandic
% This shows step and initial condition responses for a
% system with three poles.
clf
[a,b,c,d]=zp2ss([-2],[-1     -5+20*i      -5-20*i],200);
subplot(211),     step(a,b,c,d)
title('step  response')
subplot(212), initial(a,b,c,d,[0 0 1])
hold on
[y,x,t]=initial(a,b,c,d,[1 0 0]);
plot(t,y,'-.')
title('initial  condition  responses')
hold off
```

The next example returns to the system of *sonofstepvimp.m*. This time we plot the response of the system to two different inputs. This is done, in file *righton.m*, by generating input signals **u** and **unew**. We then use the

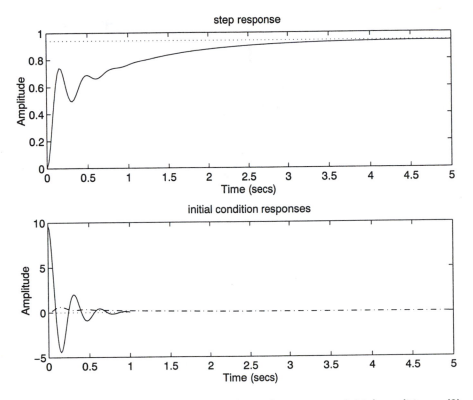

Figure 4.8   Step response in upper plot and response to initial conditions $x(0) =$ [0 0 1]$'$ (solid) and $x(0) = $ [1 0 0]$'$ (dash-dot) in lower plot for $G(s) = 200(s + 2)/(s + 1)(s + 5 - 20i)\,(s + 5 + 20i)$.

command **lsim** to calculate and plot the responses. The plots are shown in Figure 4.9. Notice that we superposed the input signals on the plots by means of the **hold on** and **plot** commands. The inputs were chosen to correspond, except for amplitude, to the inverse Laplace transform of the real pole of the system (in the first plot) and to the inverse Laplace transform of the complex conjugate pair of poles taken together (in the second plot). Notice how much longer and larger the initial transient response is in the second plot.

righton.m

```
% righton
% This shows what happens when the inputs are
% right on the poles.
clf
[num,den]=zp2tf([-2],[-1    -0.2+10*i    -0.2-10*i],100);
```

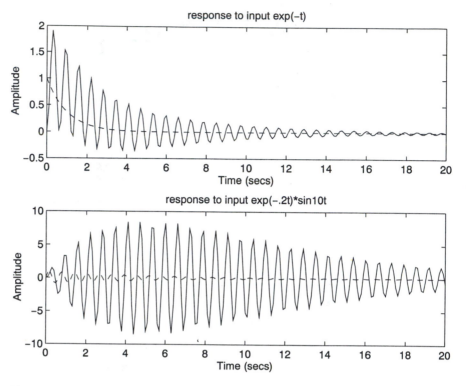

Figure 4.9   Response   of $G(s) = 100(s+2)/(s+1)(s+0.2+10i)(s+0.2-10i)$ to inputs $u(t) = e^{-t}$ and $u(t) = e^{-0.2t}\sin 10t$ for $t \geq 0$. Inputs are shown dashed.

```
t=[0:.1:20];
u=exp(-t);
subplot(211),     lsim(num,den,u,t)
hold on
plot(t,u,'m- -')     % This shows the input as a magenta
                     % dashed line on color monitors
title('response   to   input   exp(-t)')
hold off
unew=exp(-0.2*t).*sin(10*t);
subplot(212),     lsim(num,den,unew,t)
hold on
plot(t,unew,'m- -')
title('response   to   input   exp(-0.2*t)sin(10t)')
hold off
```

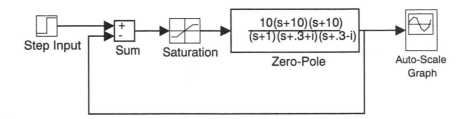

Figure 4.10 SIMULINK diagram for *constabex.m*, an example of a conditionally stable system.

4.4 Time Response of Nonlinear Systems Using SIMULINK

SIMULINK makes it easy to compute time responses for a variety of nonlinear systems. We used SIMULINK to create Figures 4.10–4.12 as an example. Notice in Figure 4.10 that the nonlinear system we created, exclusive of the **Step Input** and the **Auto-Scale Graph**, is a feedback system with a **Saturation** nonlinearity. Such systems are extremely common because most actuators ultimately saturate. In many applications it is too expensive or otherwise impractical to avoid saturation.

This system was chosen because it is simple and its step response is interesting. We used the default version of the **saturation** block from the directory named **Nonlinear**. We also used the **Sum** block from the **Linear** directory, double-clicked on the **Sum** icon to open it, and changed the signs to + –. **Zero-pole** was also found in **Linear**. We double-clicked to open it and then set the poles, zeros, and gain as shown in Figure 4.10.

The **Step Input** is from the collection of **Sources**. Its default value was used to create Figure 4.11. We double-clicked on the **Step Input** icon to open it and set the **Step Amplitude** to 10 to produce Figure 4.12. We find **Auto-Scale Graph** the easiest to use of the various display blocks in the **Sinks** directory. We used the default version to produce both Figures 4.11 and 4.12.

Finally, the **Simulation** menu was pulled down to **Parameters**. We used the default values of the parameters with the exception of **Stop Time**. This was set to 4 seconds to produce Figure 4.11 and to 40 seconds to produce Figure 4.12.

Why is the response of this system to a unit step so good while the response to a step of amplitude 10 is so bad? As will be seen at the end of Chapter 5, this system is an example of a conditionally stable system. That

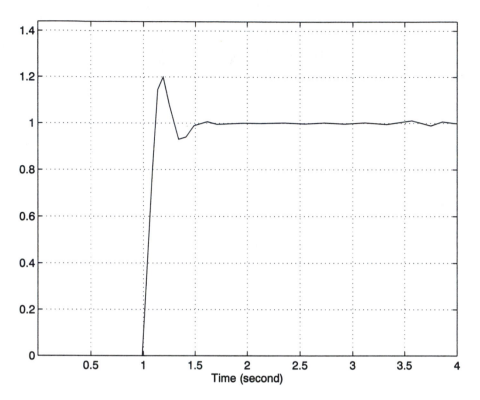

Figure 4.11   Response of the system of Figure 4.10, *constabex.m*, to a step of amplitude 1 applied at *t* = 1.

is, the closed-loop system is stable for a large gain (the gain of 10 in **Zero-Pole**, for example) but not for some small gains. It is very common for such systems to become unstable when their input signal is too large. For this reason, engineers avoid designing conditionally stable systems whenever possible.

It is interesting to replace the saturation nonlinearity in our example by other nonlinearities and see if similar results hold. For example, try replacing **Saturation** by **Relay**, **Dead Zone**, **Coulomb Friction**, and some of the other built-in nonlinearities.

Time responses for nonlinear systems can also be produced without using SIMULINK. The relevant MATLAB commands are ordinary differential equation solvers called **ode23** and **ode45**. These commands are particularly useful when your nonlinear system is described as state equations of the form $\dot{x} = f(x, t)$. In fact, in this case these commands may be easier to use than SIMULINK because putting the system into block diagram form can be cumbersome. To use **ode23** or **ode45**, you first create a function,

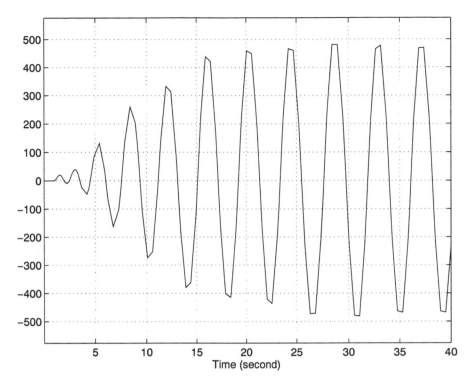

Figure 4.12 Response of the system of Figure 4.10, *constabex.m*, to a step of amplitude 10 applied at $t = 1$.

e.g., **xdot**, that computes $f(x, t)$. Then, you compute the time response using **ode23** or **ode45** with **'xdot'** as an input argument. Type **help ode45** for details.

4.5 Exercises

1. Construct a sixth-order system; compute its poles and zeros; and display the pole-zero map. All the coefficients must lie in the interval [0,10]. Answer the following two questions empirically by changing the coefficients. What is the longest distance achievable between a pair of poles? What is the longest distance achievable between a zero and a pole?

2. (a) Construct ten poles as follows. Pole number one is at +1. The poles are equally spaced around the unit circle in the complex plane. Now construct a transfer function with no zeros and the five poles, out

of the ten, that are in the left half of the complex plane.

(b) Compute and plot the step response of the system you created in part (a). What are the time to peak, percent overshoot, rise time, and settling time for this step response?

3. Repeat Exercise 2 for systems with 6 and 14 poles. Compare the results. If n equals the number of poles of the system, do you see a trend as n increases?

4. Add a zero to the system in *meshplot.m*. Generate the mesh plot for this new system and compare it to the one in Figure 4.3.

5. Generate a family of 12 second-order systems such that

(a) dc gain is constant

(b) there are no zeros

(c) the poles have real part equal to –1

(d) the poles have imaginary part varying from 0 to 4

Plot the mesh plot of the step responses of the 12 systems.

6. Repeat Exercise 5 with a zero at –2.

7. Generate the mesh plot of the step responses for 12 second-order systems having

(a) dc gain constant

(b) no zeros

(c) the poles varying from +1,+1 to –1,–1 along a circle of radius 1.

Hint: Use **ord2** and keep ω_n constant.

8. Repeat Exercise 7 but move the poles along lines of constant ζ:

(a) $\zeta = 0.1$

(b) $\zeta = 0.707$

9. Modify *approx.m* by replacing the pole at $-1-n$ by a pair of poles at $-5 \pm ni$. Run the new file to see how closely a second-order system can approximate a fourth-order system with complex conjugate poles.

5 ROOT LOCUS PLOTS

MATLAB makes it very easy to create root locus plots and extract information from them. Just a few minutes of effort can produce whole families of root locus plots. These can greatly enhance your understanding of linear systems and controller design. Thus, although this chapter begins by describing and illustrating the creation of root locus plots, the main emphasis is on how to use these plots to learn about, and to design, control systems. The important new commands introduced in this chapter are **rlocus**, **rlocfind**, and **sgrid**.

5.1 Creating and Analyzing Root Locus Plots

Root locus plots illustrate how the closed-loop poles of a system change as a system parameter is varied. These plots are particularly useful for control system design when the plant can be modeled by a rational transfer function. Creating root locus plots in MATLAB is simplicity itself. The standard feedback control system for analysis by root locus is shown in Figure 5.1. Suppose the plant to be controlled is described by the transfer function

$$G(s) = \frac{s + 2}{s^2 + 6s + 10}$$

The variable gain is denoted by k. Create the file *simplerootlocus.m*.

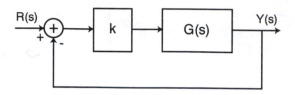

Figure 5.1   Unity feedback control system.

simplerootlocus.m

```
% simplerootlocus
% This creates a root locus plot for
% G(s)=(s+2)/s^2+6s+10).
clf
num=[1 2];
den=[1 6 10];
rlocus(num,den);
```

Running the file produces Figure 5.2, the root locus plot of $G(s)$.

This example was chosen because it is simple and has an interesting-looking plot. The plant has a zero at –2 and a pair of complex conjugate poles at $-3 + i$ and $-3 - i$. Notice that these can be read from the graph. The commands that create the plot are not new, with the exception of the last one, **rlocus**. **rlocus** calculates and plots the root locus of the system defined by the transfer function with numerator **num** and denominator **den** for a vector of values for k that is automatically determined.

The second example demonstrates that the root locus can be computed for a plant given in state-space form. The MATLAB commands are given in file *ssrloc.m*. Notice that the plant in this example has four poles.

ssrloc.m

```
% ssrloc
% This illustrates the use of rlocus with a system
% in state space form.
clf
a =      [-1.5    -13.5    -13.0    0
          1        0        0       0
          0        1        0       0
          0        0        1       0];
b =      [1; 0; 0; 0];
```

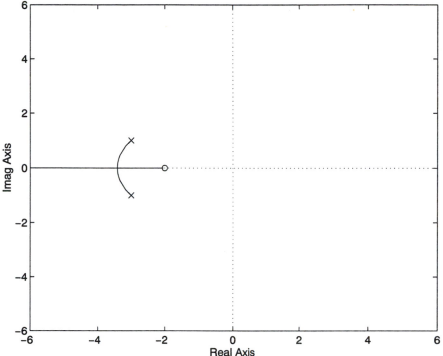

Figure 5.2 Root locus plot for *simplerootlocus.m*.

```
c =      [0 0 0 1];
d =      0;
rlocus(a,b,c,d);
```

MATLAB recognizes that the problem is in state-space form from the arguments of the command **rlocus**. The root locus is shown in Figure 5.3. Notice that we have uncovered a minor flaw in MATLAB. The corners and straight lines in the portion of the locus that is off the real axis are slightly inaccurate. It just doesn't look as nice as it should. We will show you how to fix it shortly.

If you actually execute *ssrolc.m* you will get the following message from MATLAB.

Warning: Divide by zero.

Such error messages can be important and should not be ignored. However, in this case we know that the points on the root locus are computed by solving for the eigenvalues of a 4×4 matrix. This should be a trouble-free calculation. Furthermore, our understanding of the root locus theory

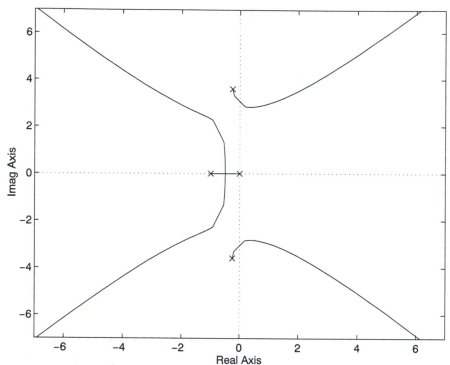

Figure 5.3   Root locus plot for *ssrloc.m*.

and our previous experience with root loci suggest that the plot in Figure 5.3 is correct. Thus, we conclude that the warning is probably due to the open-loop pole at the origin and can be safely ignored. If we felt the like-lihood of a genuine problem were higher, we could try to locate the exact source of the warning message.

This example illustrates a fairly common control problem. The two poles on the real axis are typical of many mechanical actuators. The complex conjugate pair of very lightly damped poles are typical of flexible struc-tures. You might think of the system as a simplified version of the lin-earized pitch axis transfer function of an airplane. The two pure real poles are the rigid body dynamics. The lightly damped pair of complex conjugate poles describe the flexure of the wings. Real aircraft transfer functions have more poles, including additional poles that correspond to wing flexure.

As you can see from the root locus, this is a fairly challenging control problem. Increasing the gain, k, soon causes instability. Even at fairly low gains, the lightly damped poles have an objectionable effect. As shown in the previous chapter, they will add a high-frequency "ring" to the transient response of the closed-loop system.

It is possible to extract much more information from the root locus plots than we have so far. For example, it would be nice to know the exact value of *k* for which the closed-loop system in *ssrloc.m* becomes unstable. As we will explain shortly, it is better to work with transfer functions than state-space models. Thus, we demonstrate the ideas using the transfer function generated in *simplerootlocus.m*. After executing *simplerootlocus.m* so as to display the root locus on the computer screen, type

```
>>[k,poles] = rlocfind(num,den)
```

to obtain a crosshair cursor which you can move, by means of the mouse (or arrow keys on some computers), to any point on the root locus. Moving it to any point on the locus and clicking the mouse (or hitting a key on some computers) causes MATLAB to display the value of *k* that places a closed-loop pole at that point and the location of all closed-loop poles corresponding to that value of *k*. Of course, clicking at a different location on the locus will display that value of *k* and the location of the corresponding closed-loop poles. Note that you may not be able to click exactly where you want. The computer maps the cursor position onto a rather coarse grid on its screen.

What if you click at some random point, far from the root locus? MATLAB does something rather odd. Given an arbitrary point *p* in the complex plane—the point where you clicked the mouse—MATLAB computes a corresponding value for *k*,

$$k = -\mathrm{den}(p)/\mathrm{num}(p)$$

where $\mathrm{num}(p)$ denotes the value of the numerator polynomial of $G(s)$ evaluated at *p* and $\mathrm{den}(p)$ denotes the value of the denominator.

If *p* is on the root locus, *k* will be a real positive number. However, as long as *p* is a complex number that is not on the root locus, the resulting value of *k* will be a complex number. Alternatively, if *p* is a real number that is not on the root locus, then the resulting *k* will be negative. MATLAB then computes the modulus of *k* (absolute value if *k* is real) and returns this to you as the value of *k* corresponding to the point you chose. It then computes the pole locations on the actual root locus that correspond to this real value of *k*. Therefore, the value of *k* and corresponding poles returned by MATLAB will be on the root locus no matter where you click the mouse. To illustrate this for *simplerootlocus.m*, we used the mouse to place the crosshair close to the point $0 + 4i$, which is not near the root locus, and clicked. MATLAB returned a gain **k** and the corresponding pole location **poles** on the root locus as follows.

```
selected_point =
    -0.0042 + 4.002i
k =
    5.5333

poles =
    -9.2577
    -2.2756
```

There is another way to use **rlocfind**. Suppose you do not want to bother with the mouse. Instead you read a value of *p* directly off the root locus by inspection. An obvious candidate for *simplerootlocus.m* is the breakaway point on the locus shown in Figure 5.2. It appears to be at $s = -3.4$. If you type

```
>>[k,poles]=rlocfind(num,den,-3.4)
```

MATLAB will return

```
k =
    0.8286
poles =
    -3.4286
    -3.4000
```

The corresponding **rlocfind** commands for a state-space model description as in *ssrloc.m* would be

```
>> [k,poles]=rlocfind(a,b,c,d,p)
```

where **p** is any complex number you choose—preferably one close to the actual root locus, for the reasons outlined in the previous paragraph. In fact, **p** can be a vector of complex numbers. MATLAB is then supposed to return the corresponding vector of values of **k** and the matrix of corresponding poles.

We advise against using **rlocfind** in its state-space version. The reason is that the state-space version of **rlocfind** uses the numerically unstable **ss2tf** to transform the state-space system to transfer function form. The transfer function is then used to determine the gain. If the transfer function is wrong, then the results returned by **rlocfind** will be wrong. The command **ss2tf** does not give a warning message.

To illustrate the problem, we executed *ssrloc.m* and then entered

```
>> [k,poles]=rlocfind(a,b,c,d)
```

We clicked the mouse at the point where the root locus crosses the $j\omega$-axis as you can see from the MATLAB response for the following selected point. The values for the poles are wrong, as is evident from the MATLAB response. The value for k is also wrong, but that is less obvious.

Select a point in the graphics window

```
selected_point =
     0.0354 + 3.0064i
k =
     0.7400
poles =
     -0.2489 + 3.5891i
     -0.2489 - 3.5891i
     -0.9414
     -0.0607
```

The reason for the wrong answer can be found by entering

```
>> [num,den] = ss2tf(a,b,c,d)
>> roots(num)
```

What has happened is that a small inaccuracy in computing **num** has introduced a set of incorrect zeros in the transfer function. These zeros cause the computed value of k to be wrong.

If you want to use **rlocfind** on a state-space system we suggest that you first convert the system to transfer function form, carefully check the transfer function to be sure it is correct, correct it if it is not, and then do the root locus calculations with the transfer function.

5.2 Customizing Root Locus Plots

There are many reasons for customizing a root locus plot. The file *noseeum.m* demonstrates one such reason.

noseeum.m
```
% noseeum
% Example of a root locus that is hard to see.
clf
num=1;
den=conv([1  20], conv([1  0], [1  1]));
rlocus(num,den)
```

The example system has no zeros and three poles, at 0, –1, and –20. It is a good example of an open-loop plant that is predominantly second order. Many engineers would ignore the pole at –20 and simply plot the root locus for the other two poles. With MATLAB, there is not much point in doing that because you can produce the root locus for three poles almost as quickly as for two. The example is fairly realistic. An electric motor rotating a damped inertial load typically has a pair of poles that correspond to the transfer function from torque to shaft angle. One of these poles is at 0 and the other is at a fairly small negative real number which we can always choose to be one by appropriate normalization. The transfer function from volts to torque is usually much faster. The pole at –20 is a reasonable approximation to this transfer function. The overall transfer function is just the product of these two.

The root locus that results from running *noseeum.m* is shown in Figure 5.4. It is difficult to see the important part of the root locus. We don't care about the pole at –20 as it proceeds toward $-\infty$. We only care about the two poles that are close to the $j\omega$-axis. However, we do care about the effect that the pole at –20 has on the root locus, so we do not drop it from our model. Instead we customize the graph in order to see more clearly the portion of the root locus that corresponds to the dominant poles. The result is file *seeum.m*.

seeum.m
```
% seeum
% Example of a customized root locus.
clf
num=1;
```

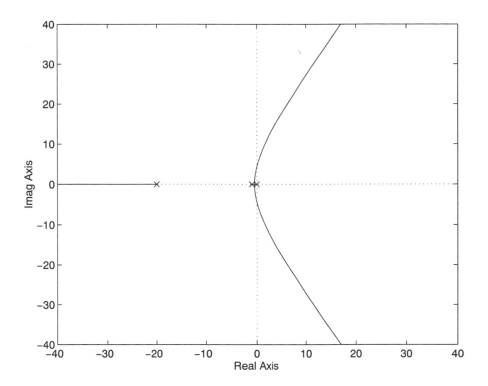

Figure 5.4 Root locus for *noseeum.m*.

```
den=conv([1    20],conv([1    0], [1    1]));
rlocus(num,den)
axis([-2 2 -6 6])
```

This is just *noseeum.m* except for the **axis** command, which is used to plot the root locus with the desired formatting as shown in Figure 5.5. Notice that the uninteresting part of the root locus is not visible. It is about 8 feet to the left of the paper. However, what you see does reflect the influence of the extra pole. Whereas the root locus for a system with two poles never crosses the $j\omega$-axis for positive k, the root locus with the three poles of the system of this example does cross the $j\omega$-axis for a sufficiently high value of k, thereby indicating an unstable closed-loop system.

The previous examples have demonstrated that MATLAB is a very effective tool for creating root locus plots. However, it is not foolproof.

We created the following example, *whoops.m*, to demonstrate that MATLAB does not always produce correct root locus plots.

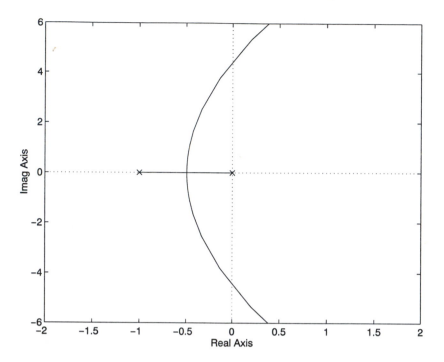

Figure 5.5   Root locus for *seeum.m.*

<u>*whoops.m*</u>
```
% whoops
% Example where MATLAB produces an incorrect root
% locus plot.
num=[1 1 1.3];
den=conv([1 0],conv([1 1],conv([1 1 1.25],[1 1 4.25])));
rlocus(num,den)
axis([-1 0 .5 1.5])
```

The result of executing *whoops.m* is shown in Figure 5.6. A simple theoretical calculation shows that this root locus should be symmetrical about the vertical line $s = -0.5$. The actual error is quite small and you easily correct it with your eyes. But, it is an aesthetically unpleasant and glaring error, and it is easy to fix.

Having executed *whoops.m* to define **num** and **den**, you can then execute *fixit.m*, which follows, to produce a better root locus plot for $G(s) = (s^2 + s + 1.3)/s(s + 1)(s^2 + s + 1.25)(s^2 + s + 4.25)$.

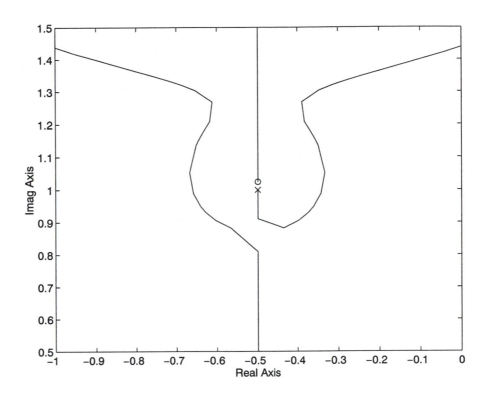

Figure 5.6 Root locus plot produced by *whoops.m*. The plot should be symmetric about -0.5.

fixit.m
```
% fixit
% This fixes the root locus plot produced by whoops.m.
clf
[p,k]=rlocus(num,den);
rlocus(num,den,k')
axis([-1 0 .5 1.5])
[kstar,pstar]=rlocfind(num,den)
l=(k<kstar);
index=sum(l);
kadd=[k(index):.001:k(index+1)];
knew=[k(1:index)' kadd k(index+1:length(k))'];
clf
rlocusnew(num,den,knew)
axis([-1 0 .5 1.5])
```

The first line of *fixit.m* after the comments gets rid of the root locus plot created by *whoops.m*. The second line creates a column vector of the values of the gain, **k**, for which *whoops.m* computed points on the root locus. The next two lines actually recreate the root locus plot of Figure 5.6 except that the connecting lines are omitted. We then use **rlocfind** as a convenient means of identifying the region of the root locus plot where we need more points. We clicked the mouse at the last point before the breakaway from the $s = -0.5$ vertical line to obtain the MATLAB response.

```
selected_point =
     -0.5000+0.8109i
kstar =
     2.6470
pstar =
     -0.5000+1.8064i
     -0.5000-1.8064i
     -0.5000+0.9106i
     -0.5000-0.9106i
     -0.5000+0.8109i
     -0.5000-0.8109i
```

The next three lines of *fixit.m* create a new vector of gains, **knew**, at which to evaluate the root locus. This new vector contains many extra points in the region where the old plot was incorrect. The final three lines get rid of the old plot and create the new one shown in Figure 5.7.

We need to explain **rlocusnew**. If you supply the vector of gains at which the root locus is to be evaluated to the current version of **rlocus**, then the points on the root locus are not connected. Instead, little crosses appear at the pole locations. In order to get lines connecting the points and get rid of the crosses, we made a copy of *rlocus.m*, edited it, and saved it as *rlocusnew.m*. We made four changes. We replaced **rlocus** by **rlocusnew** in the first line. We added a comment line explaining the changes we made. We changed + to – in the **plot** commands in the section of *rlocus.m* that plots the root locus when the gains are given.

You might think we are being excessively picky about the exact shape of the root locus. Perhaps we are. The fact remains that MATLAB occasionally gives slightly erroneous root loci. There are examples among the Exercises. The problem is that MATLAB attempts to construct the *k* vector based on interpolation between computed roots and a given percent error threshold. However, this threshold may not be appropriate for all

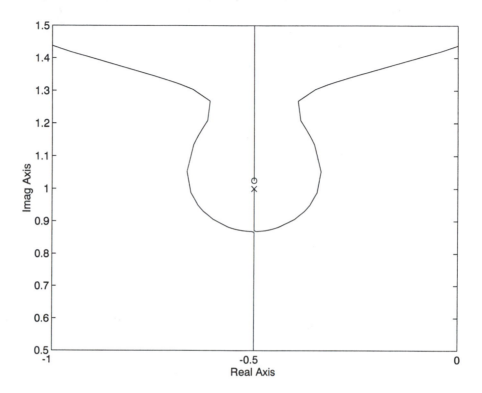

Figure 5.7 Improved root locus plot for $G(s) = (s^2 + s + 1.3)/s(s + 1)$ $(s^2 + s + 1.25)(s^2 + s + 4.25)$.

root locus plots. While the corresponding points on the locus are accurately computed, the curves connecting these points are computed by a simple interpolation procedure and are not as accurate. This accounts for the ugly curves we have already seen. As an alternative to *fixit.m* you can modify the **rlocus** function itself. To do this, first copy the file *rlocus.m* to a new file, e.g., *rlocusnew.m*. Edit *rlocusnew.m* to increase the value of the variable **precision** from 1 to 10 (or higher). Next repeat *whoops.m*, replacing **rlocus** with **rlocusnew**.

A more serious problem can occur when two or more branches of the root locus are near each other. MATLAB will sometimes connect the wrong points. The resulting plot usually looks strange. However, it is possible that a plot that looks correct is not. Be especially cautious about root locus plots containing branches that are close to each other. If there are doubts, the procedure illustrated for customizing root locus plots should

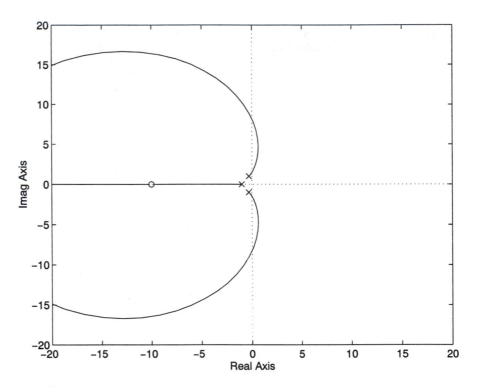

Figure 5.8   Root locus plot for $G(s) = 20(s + 10)^2/(s + 1)(s + 0.3 + i)(s + 0.3 - i)$.

be used to produce a more accurate plot in the questionable region.

As a last example of plotting root loci we used *conditstab.m* to plot the root locus for the SIMULINK example *constabex.m* from Figure 4.10.

conditstab.m

```
% conditstab
% Example of a conditionally stable system.
clf
num=20*conv([1 10],[1 10]);
den=conv([1 1], conv([1 .3+i],[1 .3-i]));
rlocus(num,den)
```

The saturation nonlinearity is ignored. The resulting root locus is shown in Figure 5.8. Notice that the closed-loop system goes from stable to unstable and then back to stable as the feedback gain increases from 0 to ∞. Such systems are called conditionally stable systems. Control system engineers try very hard to avoid designing conditionally stable systems because of the behavior illustrated in Figures 4.11 and 4.12.

5.3 Design by Means of Root Locus Plots

The root locus plot can be an extremely useful design tool. We illustrate its use in controller design here. While we are at it, we also demonstrate several other useful MATLAB features.

Design example: Suppose we have a plant $G(s) = 1/s(s + 1)(0.2s + 1)$ that we would like to control. Notice that we have normalized the plant description so that $\lim_{s \to 0} sG(s) = 1$. This will make it easier to compare different controller designs.

The simplest controller is a pure gain, so we first try to design the best possible controller using only a gain k. Of course, we need to agree on the specifications, the rules by which we will decide which controller is best. We will require our closed-loop system to have a damping ratio of exactly 0.707 and the largest possible natural frequency. These are realistic design goals except that we have made them somewhat less rigid than normal in order to facilitate the comparison of different candidate designs.

The first step in arriving at a design is to plot the root locus. Running file *rlocdesignex.m* will produce the root locus shown as the line leaving the real axis near $s = -0.5$ in Figure 5.9. We have expanded the plot in the region of major interest so you can see better. Because the root locus is symmetric about the real axis you don't really need the third and fourth quadrants. We would not design an unstable system, so the first quadrant is unnecessary. The pole at -5 is simply going to decrease to $-\infty$. We have also introduced a new tool. The command **sgrid** has been used to superimpose both a 45° line, corresponding to a damping ratio of 0.707, and dotted curves, corresponding to natural frequencies $\omega_n = 1, 2, 3$, and 4, on the plot.

rlocdesignex.m

```
% rlocdesignex
% A textbook example of design by means of the root
% locus.
clf
num=1;
den=conv(conv([1 0], [1 1]), [0.2 1]);
rlocus(num,den);
v1=0.1; v2=2.5; h1=4; h2=0.1;
```

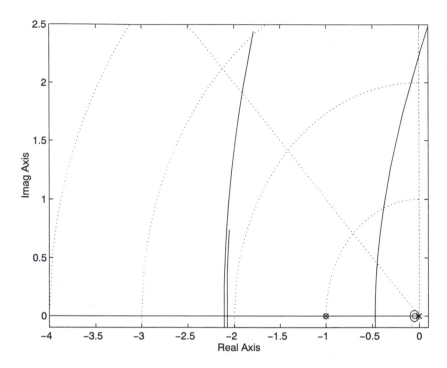

Figure 5.9   Root locus for *G(s)= 1/s(s+ 1)(0.2s+ 1)* with constant-gain controller (line leaving the real axis near –0.5), lead compensator (long line leaving the real axis near –2), and lead-lag compensator (short line leaving the real axis near –2).

```
axis([-h1 h2 -v1 v2]);
damping=0.707;
wn=1:1:4;
sgrid(damping,wn)
```

We immediately see that the pure gain controller that achieves a damping coefficient of 0.707 would have a natural frequency of roughly 1/2. The corresponding gain is approximately $k = 0.41$. The value was found using the command **[k,poles]=rlocfind(num,den)** and clicking on the graph at the point where the locus crosses the line of 0.707 damping. You can check the result by looking at the poles of the closed-loop system with the **cloop** and **roots** commands as follows:

```
>>[numcl,dencl]=cloop(0.41*num,den);
>>roots(dencl)
```

You will see that the poles of the closed-loop system really do correspond to the selected point and you will also be able to check that the damping ratio of the dominant poles is about 0.707. The corresponding natural frequency is computed to be 0.64.

The command **sgrid** provides a useful tool when designing and analyzing control systems using root locus plots. Without any arguments **sgrid** superimposes a grid on an existing root locus plot (or pole-zero map) with ten lines of constant damping ratio equally spaced from 0.1 to 1.0 and lines of constant natural frequency ω_n. However, **sgrid** can also be used with arguments for a customized grid as **sgrid(dmp,freq)** where **dmp** is a vector of damping ratios and **freq** is a vector of natural frequencies to be plotted. See *rlocdesignex.m* for an example.

The command **sgrid** can also be called as **sgrid('new')** or **sgrid(dmp,freq, 'new')**, which will clear the screen before plotting the grid and set **hold on** for the next plot commands.

Adding a lead compensator would be a standard way to improve the natural frequency of the closed-loop system of *rlocusdesignex.m*. For a system with three real poles one normally puts the zero of the compensator close to the middle pole of the plant. The compensator pole is placed as far to the left as possible. Generally, noise and actuator limits prevent one from making this pole larger than ten times the compensator zero. We have chosen the lead compensator $C(s) = k(s + 1)/(0.1s + 1)$. Notice that $\lim_{s \to 0} sC(s)G(s) = k$. The corresponding root locus is created and superimposed on the previous example by entering *addlead.m* immediately after entering *rlocusdesignex.m*. It is shown as the long line leaving the real axis near $s=-2$ in Figure 5.9. This is a great improvement. Choosing the controller gain that produces a damping ratio of 0.707 also produces a natural frequency of roughly 2.7. The corresponding gain is approximately $k = 1.64$ computed and checked as for *rlocusdesignex.m*.

addlead.m

```
% addlead
% This adds a lead compensator to the
% previous example.
hold on
num=[1 1];
den=conv(conv(conv([1 0], [1 1]), [0.2 1]), [0.1 1]);
rlocus(num,den)
hold off
```

Having done such a good job, suppose we are asked if we can improve the steady-state performance of the closed-loop system without signif-

icantly reducing the natural frequency. The standard way to do this is
by adding a lag compensator. Again, it is generally not feasible to have
the compensator pole and zero more than a factor of ten apart. The
pole should be close to the origin. We have chosen a pole at –0.005
and a zero at –0.05. Combined with the lead compensator this makes
$C(s) = k(s + 1)(20s + 1)/(0.1s + 1)(200s + 1)$. The corresponding root
locus is shown as the solid curve in Figure 5.9. It is produced by running
file *addlag.m* (not shown because it is very similar to its predecessors)
immediately after the previous two files. Notice that the lag compensator
has decreased the natural frequency only slightly. There is an extra pole
and it is close to the origin, but it has almost no effect on the transient
response because of the zero that is nearby on the right.

On the other hand, the lag compensator has improved system performance by decreasing the steady-state error. We find that the gain corresponding to a damping ratio of 0.707 is now approximately $k = 16.35$.
Therefore, with both the lead and lag compensator the steady-state error
for a unit ramp input is $1/\lim_{s \to 0} sC(s)G(s) = 1/k = 0.06$. With the lead
compensator only, the appropriate gain was found to be $k = 1.64$. This
means the steady-state error would be 0.6. The advantage of the lag compensator is illustrated by comparing these steady-state error results. The
gain for the lead plus lag compensator was computed as before. The value
of k was checked similarly to the previous two examples. It is important
to realize that the gain, k, must be fairly large for a lag compensator to
work properly. If the gain is not large enough to push the compensator
pole close to the compensator zero, the lag compensator will damage the
transient response.

5.4 Simulation of PID Control Using SIMULINK

One of the most common industrial control systems is the PID controller.
Theoretically, the analysis and design of PID controls are similar to the
analysis and design of lead-lag controls. However, the physical implemen-
tations of PID controllers are notorious for a problem known as integral
windup that results from a nonlinear effect, actuator saturation. We use
SIMULINK to study integral windup and its prevention.

SIMULINK includes both a textbook PID controller and a sophisticated PID
controller in the **extras** directory under **PID Controllers**. The textbook
version is called **PID Controller**, while the more sophisticated version

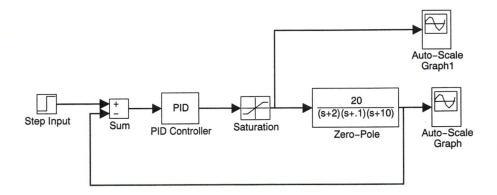

Figure 5.10   A simple example, *piddle.m*, of a plant with a saturating actuator and a textbook PID controller.

is called **Set point PID with Anti-Windup**. As you might guess, the sophisticated PID controller includes some components to prevent integral windup.

The first step in this introduction to the realities of PID control is to use SIMULINK to create *piddle.m* as shown in Figure 5.10. Set the **Sum** and **Zero-Pole** blocks as shown. Set the lower output limit of the **Saturation** block to –1 and the upper output limit to 1. The next step is to choose the gains of the PID controller.

There are many ways to determine the appropriate gains for a PID controller. In fact, many industrial PID controllers come equipped with three knobs with which the control engineer is expected to dial in the proper gains. In the textbook version, the PID controller is

$$C(s) = k_p + \frac{k_i}{s} + k_d s \tag{5.1}$$

and the gains are k_p, k_i, and k_d. Of course, the $k_d s$ part of the controller in (5.1) is not really implementable. We cannot actually build systems with more zeros than poles. There is another problem. PID controllers are usually placed in systems that have step inputs, exactly as in *piddle.m*. When a step input occurs while the system is at rest the D term in the PID controller differentiates the step and creates an impulse. This is usually undesirable.

For the reasons given above, and to simplify the example somewhat, we set $k_d = 0$. This changes the controller in (5.1) into a PI controller. Such a controller is basically a lag controller with its pole at the origin. We

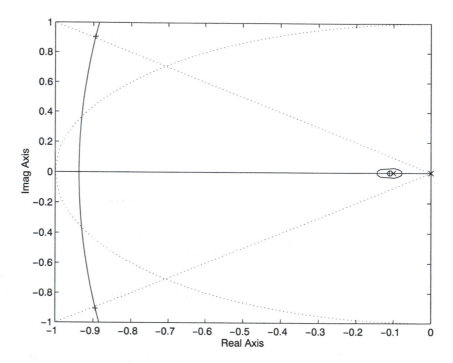

Figure 5.11 Root locus plot for the design of a PI controller for $G(s) = 20/(s + 0.1)(s + 2)(s + 10)$.

then placed the controller zero close to the system pole at –0.1 and used MATLAB to create *designpi.m* to find the correct gain for a damping ratio of 0.707. We then executed *designpi.m*, clicked the mouse at the point in Figure 5.11 where the locus crosses the 0.707 damping ratio line, and obtained k_p = 0.8370 and k_i = 0.0921. We put these values into the **PID controller** block for **Proportional** and **Integral** respectively and 0 for **Derivative**.

We then pulled down the **Simulation** menu to **Parameters...** and set the **Stop Time** to 15. We used the default values for the rest of the simulation parameters. Finally, we set the **Final Value** of the **Step** block to 0.5. We then ran the simulation to produce Figure 5.12. The figure shows the output of the closed-loop system when there is no integral windup. We then changed the **Final Value** of the **Step** block to 5 and ran the simulation again. The resulting output is shown in Figure 5.13. You can see the effect of integral windup by comparing the two responses. The saturation causes the integral of the error to accumulate to a much larger value than in the linear case. This large integrated error, known as integral windup for historical reasons, causes the large percentage overshoot and the long

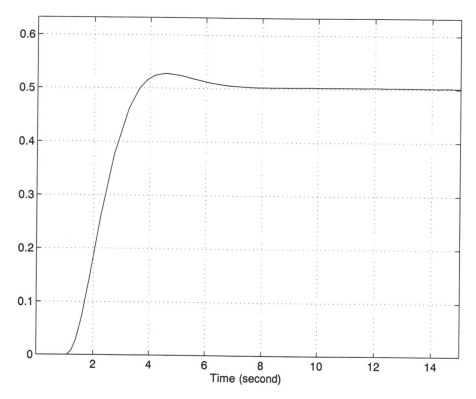

Figure 5.12   Response of the system in Figure 5.10 to a step of amplitude 0.5 applied at *t* = 1. The closed-loop system behaves linearly.

settling time seen in Figure 5.13. **Auto-Scale Graph1** displays the saturation when the input step has final value 5 and the absence of saturation when the input step has final value 0.5.

designpi.m

```
% designpi
% This provides the design for a PI controller for the
% plant G(s)=20/(s+10)(s+.1)(s+2).
clear
num=20*[1 .11];
den=conv([1 10],conv([1 .1],conv([1 0],[1 2])));
rlocus(num,den)
sgrid([.707],[1])
axis([-1 0 -1 1])
[kp,poles]=rlocfind(num,den)
ki=0.11*kp
```

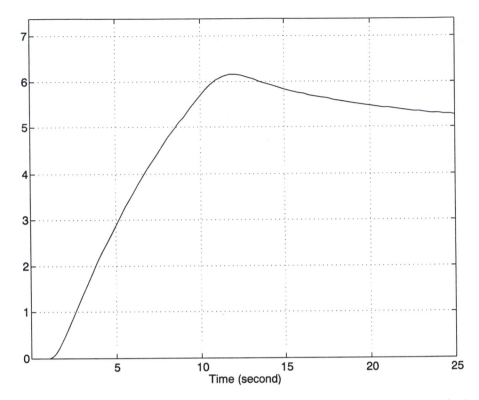

Figure 5.13   Response of the system in Figure 5.10 to a step of amplitude 5 applied at $t=1$. This response exhibits the effect of integral windup.

Now that we have seen the evils of integral windup, what can be done about it? One approach is to use a PID controller that has some form of anti-windup system incorporated within it. To see what such a controller can do, edit *piddle.m* to create *impiddle.m* as shown in Figure 5.14. Set the **Final Value** of the **Step** block to 5. Open the **Set point PID with Anti-Windup** block and set the **Gain K** to 0.8370, **Setpoint b** to 1, **Integral Ti** to 0.8370/0.0921, **Antiwindup gain Tt** to 0.8, **Antiwindup saturation [low,high]** to [-1,1] and **Derivative gain and divisor [Td,N]** to [0,1000]. We will explain these choices shortly. Run the simulation with these values to obtain the closed-loop response shown in Figure 5.15. Although the response is not simply ten times that shown in Figure 5.12, as it would be if the system were linear, you can see that the anti-windup has worked well.

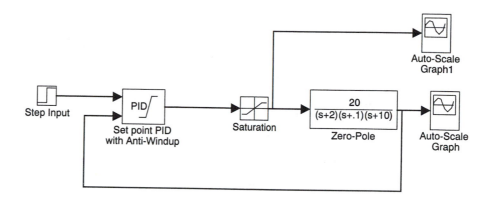

Figure 5.14 The plant of Figure 5.10 with a more realistic PID controller that includes an anti-windup device (*impiddle.m*). Note that the **sum** block is internal to the **Set point PID with Anti-Windup**.

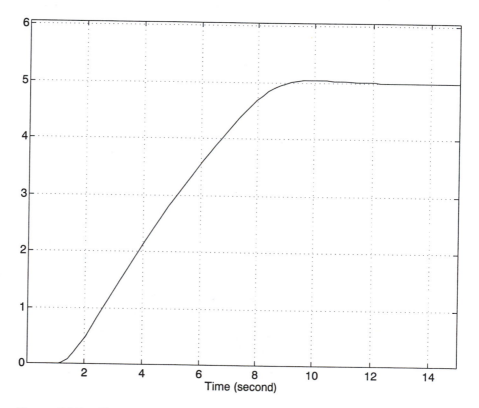

Figure 5.15 The response of the closed-loop system of Figure 5.14 to a step input of amplitude 5 at *t* = 1. Note the improvement with respect to the response in Figure 5.13.

PID controller with set point weighting and anti-windup.

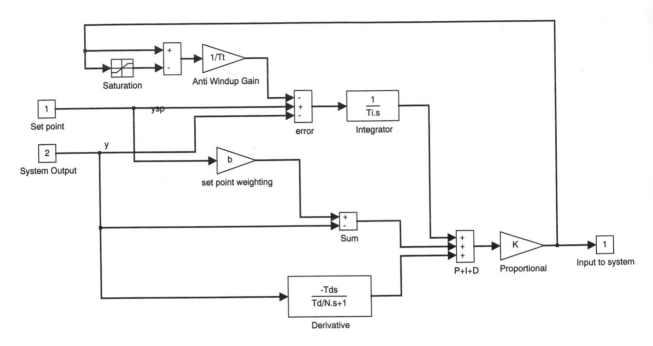

Figure 5.16   The SIMULINK **Set point PID with Anti-Windup** block unmasked.

To see how the **Set Point PID with Anti-Windup** works, open a new SIMULINK window and put only the icon for the **Set Point PID with Anti-Windup** in it. Select the icon and pull down the **Options** menu to **Unmask** and click. The appearance of the icon will change. Double-click on the icon and you will see the SIMULINK diagram for the controller shown in Figure 5.16. Notice particularly that the derivative block is realizable because it includes a pole as well as a zero. Choosing N large enough will make the implementable derivative block approximate a pure derivative at low frequencies. Also, the derivative block does not get a direct input from input 1. Thus, it does not differentiate a unit step input. Because the derivative of a unit step is zero everywhere but at the instant the step occurs, this avoids the impulse and gives the correct response everywhere except at the instant the impulse would have occurred. Of course, inputs other than steps would result in errors.

The anti-windup is accomplished by the components in the upper left corner of the diagram. Note that it is a nonlinear feedback from the output of the controller. We leave it to you to try to figure out how it works. To help you do so, see what happens if the saturation in the controller is different from the one in the system. Our choice of values made them the same. When you are finished studying the controller, pull down the **Options** menu and click on **Mask**.

Having seen the precise implementation of the **Set Point PID with Anti-Windup**, our choice of parameter values should be easily understood. Gain k is the same as gain k_p in the textbook PID controller. Integral T_i is actually k_p/k_i. By setting **td=0** we set the derivative gain to zero. We tried several values for the **Antiwindup gain Tt** before deciding that 0.8 was best.

It is very common to see industrial PID controllers with the derivative term set to zero as in our example. This is primarily because it is hard to find good values for k_d. We leave doing so for this example as an exercise. Most control textbooks contain material on designing or tuning PID controls. The book *Automatic Tuning of PID Controllers* by K.J. Åström and T. Hägglund (Instrument Society of America, 1988) contains a thorough discussion of PID control, including anti-windup.

5.5 Exercises

1. (a) Plot the root locus for $G(s) = (3s - 4)/(s^2 - 2s + 2)$.

   (b) Plot the root locus for $G(s) = -(3s - 4)/(s^2 - 2s + 2)$.

   (c) Is there any value of k that will stabilize either of these plants?

2. (a) Plot the root locus for $G(s) = (s + 2)/(s^2 + 6s + 10)$ in such a way that the real and imaginary axes are scaled identically. Hint: If you do this correctly, the complex conjugate portion of the locus will be an arc of a circle.

   (b) Now that you know that a portion of the locus is part of a circle, prove that it is.

   (c) Prove that part (b) would be true for $G(s) = (s + a)/(s^2 + 2\zeta\omega_n s + \omega_n^2)$ as long as $-1 < \zeta < 1$.

3. (a) Plot root loci for the plant $G(s) = 1/s(s+1)(s^2 + as + 4)$ as a varies from 0.01 to 4.0.

(b) See if you can find the exact value of a for which the pair of poles that goes unstable changes.

4. (a) Plot root loci for the plant $G(s) = (s^2 + s + a)/s(s + 1)(s^2 + s + 1.25)(s^2 + s + 4.25)$ as a varies from 1.25 to 4.25.

(b) See if you can find the exact value of a for which the pair of poles that asymptotically approach the zero changes.

5. Consider the root locus of *ssrloc.m*. Control designers often attempt to cancel the complex conjugate (resonant) poles by placing zeros close to them. Because they cannot build a compensator with more zeros than poles they are also forced to add poles at about the same frequency but with more damping. These are called notch filters.

(a) Add a notch filter to *ssrloc.m*—place the zeros exactly on the resonant poles. Choose the filter poles to have the same ω_n as the zeros but five times the damping coefficient.

(b) Repeat part (a) with the zeros slightly above and to the left of the resonant poles.

(c) Repeat part (a) with the zeros slightly below and to the left of the resonant poles.

(d) Given that you normally know only the approximate location of the resonant poles, which is better, (b) or (c)? Hint: The answer is determined by the stability margins. How small a change would cause your closed-loop system to be unstable?

6. Consider the design example *rlocdesignex.m*.

(a) Plot the root locus for the plant plus lead compensator only when the compensator zero is at –0.9. Remember to change the compensator pole so it is exactly ten times the zero.

(b) Repeat part (a) with the compensator zero at –1.1. Which is better?

(c) Plot root loci for the plant plus lag compensator only. How does the location of the compensator pole affect the root locus?

(d) Repeat parts (a), (b), and (c) for the plant plus lead plus lag compensators.

7. (a) Plot the root locus for $G(s) = (s^2 + s + 9.25)(s^2 + 2s + 10)(s^2 + 3s + 11.25)/s(s + 2)(s^2 + 2s + s)(s^2 + 2s + 17)$.

(b) Prove that the locus should be symmetric about the vertical line as $s = -1$.

(c) Correct the MATLAB-generated root locus.

8. (a) Plot the root locus for the system defined in *backandforth10.m*.

(b) Blow up the region near –1 by using **axis ([–1.1 –0.9 –0.1 0.1])**. Is what you see correct? Explain why MATLAB produces the result it does.

(c) Add another pole to the transfer function and repeat parts (a) and (b).

9. Find a nonzero value of **Td** that will improve the transient response of *impiddle.m*. You may also have to change the value of *N*. Finally, you will find that using **Linsim** to compute the system's response is much more efficient than using **RK-45**.

6 FREQUENCY-DOMAIN PLOTS

The designers of control systems make considerable use of frequency-domain plots. In fact, semilog plots of $20 \log|G(j\omega)|$ and $\measuredangle\, G(j\omega)$ vs. ω (Bode plots), plots of Im $G(j\omega)$ vs. Re $G(j\omega)$ (Nyquist plots), and plots of $20 \log|G(j\omega)|$ vs. $\measuredangle\, G(j\omega)$ (Nichols plots) were all standard tools in control system theory and design before Evans introduced the root locus.

Most textbooks emphasize the construction of these plots from mathematical models of linear time-invariant systems. As a result, it is easy to miss the true value and convenience of design methods based on the frequency response. All of the frequency-domain plots can be produced directly from convenient experiments on the physical system. Furthermore, it is just as easy to measure or compute the frequency response of a system with 1000 poles and zeros as it is to measure or compute the frequency response of a second-order system. Thus, design techniques that require only plots of the frequency response are much easier to apply to systems that are of very high order than either root locus or state-space techniques.

Frequency-domain plots can also be very useful when a good mathematical model exists. For example, the frequency response is much less sensitive to small errors in the model's parameters than either root locus or state-space methods. Furthermore, frequency-domain design methods are much less sensitive to the system order than state-space and root locus methods.

6.1 Creating Frequency-Domain Plots

MATLAB makes it extremely easy to create frequency-domain plots using the commands **bode, nyquist, nichols, ngrid,** and **margin.** To illustrate this for a Bode plot, we created the script file *simplebode.m.* Notice that the first two lines simply create the numerator (**num**) and denominator (**den**) polynomials for the transfer function, $G(s) = (5.9791s^3 + 53.8122s^2 + 197.3113s + 388.6435)/(s^4 + 13s^3 + 58s^2 + 306s + 260)$. The single command **bode(num,den)** then produces the Bode plot shown in Figure 6.1. Notice that this transfer function never has more than $-90°$ of phase shift.

simplebode.m
```
% simplebode
% This creates a simple Bode plot.
clf
num=[0 5.9791  53.8122  197.3113  388.6435];
den=[1  13  58  306  260];
bode(num,den)
```

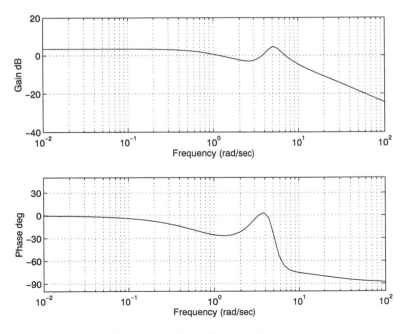

Figure 6.1   Bode plot produced by *simplebode.m.*

In *simplebode.m* we began with a transfer function and not with poles and zeros to emphasize the fact that factoring the numerator and denominator polynomials of the transfer function is difficult to do by hand. Once you know the zeros, poles, and gain of a transfer function it is actually fairly easy to create Bode plots by hand.

It generally takes more time to create Nyquist and Nichols plots than Bode plots when you do them by hand. This is true because the relation between the poles and zeros and the plot is not as simple as it is for Bode plots. MATLAB makes the creation of Nyquist and Nichols plots as easy as the creation of Bode plots. We demonstrate this with the commands **nyquist**, **nichols**, and **ngrid** in the file *simplenyqnic.m*.

simplenyqnic.m
```
% simplenyqnic
% Example of the creation of simple Nyquist and Nichols
% plots.
clf
z=[-2   -2+5*i   -2-5*i];
p=[-1   -5   -7   -1+7*i   -1-7*i];
k=(35*50)/(2*29); % k is chosen to give a dc gain of 1
[num,den]=zp2tf(z',p',k);
subplot(211), nyquist(num,den)
subplot(212), nichols(num,den)
ngrid
axis([-360 0 -40 30]);  % The default axes for Nichols
                        % plots are often inappropriate.
```

Notice that the first four lines of the file simply create the transfer function $G(s) = k(s + 2)(s + 2 + 5i)(s + 2 - 5i)/((s + 1)(s + 5)(s + 7)(s + 1 + 7i)(s + 1 - 7i))$ out of zeros, poles, and gain. We particularly wanted you to see that the system has a lightly damped pair of complex conjugate poles and a similar pair of complex conjugate zeros along with three real poles and one real zero. The Nyquist plot is produced by the command **nyquist(num,den)**. The Nichols plot could be produced by the command **nichols(num,den)** alone. However, we have generally found the Nichols plots to be uninformative unless we add the command **ngrid**. This command creates the dotted lines that allow you to read closed-loop gain and phase from the Nichols chart. We also find that it is helpful to customize the axes of the Nichols plot. This is the role of the **axis** command in the file. Note that it applies only to the Nichols plot.

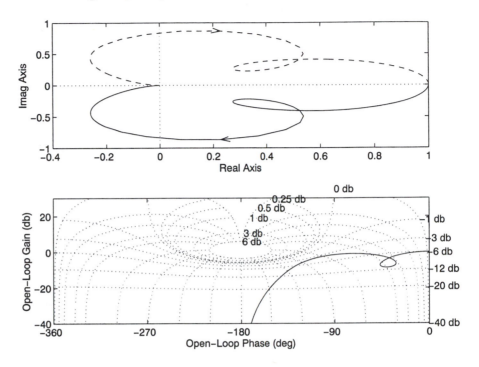

Figure 6.2   Nyquist and Nichols plots produced by file *findingsszp.m.*

Notice that the plots, shown in Figure 6.2, contain some loops. Are these reasonable? Is it possible that MATLAB has, perhaps because of numerical problems, produced erroneous plots? We can demonstrate that some aspects of the plots are reasonable and we can explain the existence of these loops. First, it is easy to verify that both plots contain a point that corresponds to $G(j0) = 1$. This is the rightmost point in both plots. Similarly, both plots show $\lim_{\omega \to \infty} |G(j\omega)| = 0$ and $\lim_{\omega \to \infty} \angle G(j\omega) = -180°$. We cut off the Nichols plot before it gets really close to the asymptote, but you can see that the asymptote is –180°.

The loops are a little harder to explain. However, since it is so easy to generate such plots we produced Figure 6.3, the Bode plot for this system, to aid us in interpreting the Nyquist and Nichols plots. Note that the zero at $-2 + 5i$ will add approximately 50° ($2 \tan^{-1} 0.5$) of positive phase shift as ω varies from 4 to 6. Other effects, notably the poles at –5 and –7, mask this slightly but you still see about 20° of positive change in phase. This is especially visible as the bottom of the loop in the Nichols plot and in the Bode plot. Next, as ω varies from 6 to 8, the pole at $-1 + 7i$ adds 90° of *negative* phase shift. This is also somewhat masked by the other poles and zeros, but it is still quite visible in all the plots.

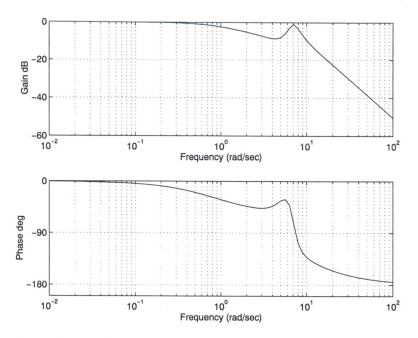

Figure 6.3   Bode plot of the system having Nyquist and Nichols plots in Figure 6.2.

The changes in magnitude are much less pronounced until $\omega \geq 7$. Notice that the phase angle of $G(j\omega)$ is close to -90° when $\omega = 7$. As you can see in the Nyquist plot, the loops are in the first and fourth quadrants. The loop in the Nichols plot occurs for phase $>-50°$. Thus, the loops in the Nyquist and Nichols plots occur for $|\omega| <7$. Checking the loop on the Nichols plot, we see that there must be two frequencies, ω, for which both $|G(j\omega)|$ and $\sphericalangle\, G(j\omega)$ are identical. In fact, $|G(j\omega_1)| = |G(j\omega_2| \approx$ -3 dB and $\sphericalangle\, G(j\omega_1) = \sphericalangle\, G(j\omega_2) \approx -40°$. Comparing these to the Bode plots gives $\omega_1 \approx 1.2$ and $\omega_2 \approx 6$. Of course, we could use MATLAB to find these values exactly. We will describe ways to obtain exact values from frequency-domain plots in the next section.

6.2 Customizing Frequency-Domain Plots

The automatic scaling routines in MATLAB can obscure important aspects of the frequency response. This is particularly true for Nyquist plots, as we will demonstrate shortly. Thus, the first issue discussed in this section is how to customize plots to display the important information.

Also, as we demonstrated in the previous section, a great deal of quantitative information is contained in the plots. Often it is important to obtain this information. The graphs are excellent media for the gestalt (the big picture), but they are not such good sources for obtaining precise numerical values as are the arrays of numbers that are used to create the graphs. Accessing these arrays is the second issue discussed in this section.

To demonstrate the potential scaling problem with Nyquist plots we created the file *nyqprob.m.*

nyqprob.m

```
% nyqprob
% Simple but poorly scaled Nyquist plot.
clf
[num,den]=zp2tf([],[-0.01  -4  -1+6*i  -1-6*i],40);
nyquist(num,den)
```

This file produces the Nyquist plot for the system described by $G(s) = 40/(s + 0.01)(s + 4)(s + 1 + 6i)(s + 1 - 6i)$ shown in Figure 6.4. Notice that there appears to be no serious stability problem. The Nyquist locus appears to be almost entirely to the right of the $j\omega$-axis. There are two warnings that indicate this might not be true, but they require careful observation. First, the scale of the plot is rather large and the Nyquist locus does appear to go slightly to the left of the $j\omega$-axis. Second, we also know that a system with four poles and no zeros should have $-180°$ of phase shift at a nonzero gain and that its Nyquist locus should be asymptotic to $-360°$ as $\omega \rightarrow \infty$. Thus, we use the **axis** command in *fixnyqprob.m* to zoom in on the origin of Figure 6.4.

fixnyqprob.m

```
% fixnyqprob
% Rescaling the Nyquist plot.
clf
[num,den]=zp2tf([],[-0.01  -4  -1+6*i  -1-6*i],40);
nyquist(num,den)
axis([-0.5  0.5  -0.5  0.5])
```

This use of the **axis** command causes the command **nyquist** to display only the 1-by-1 square range of the plot centered at the origin. The result is shown in Figure 6.5. As you can see, this system would be closed-loop unstable with a feedback gain of more than approximately 120.

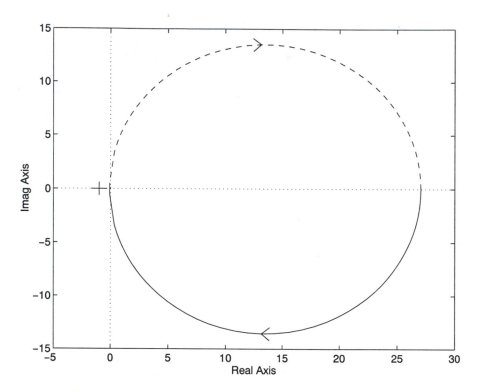

Figure 6.4 Automatically scaled Nyquist plot for $G(s)=40/(s+0.01)(s+4)$ $(s+1-6i)(s+1+6i)$.

The problem just illustrated becomes especially severe when there are poles close to the $j\omega$-axis. MATLAB produces Nyquist plots that are difficult to interpret for systems with poles on the $j\omega$-axis. Such poles introduce division by zero into the computation of Nyquist plots. In the next example, file *nyqnic.m*, we consider a system with a pole at zero,

$$G(s) = \frac{100(s + 0.1 + 5i)(s + 0.1 - 5i)}{s(s + 4)(s + 0.2 + 10i)(s + 0.2 - 10i)}$$

However, we have moved the pole from $s = 0$ to $s = -0.001$ to enable us to generate and interpret the Nyquist plot. This produces an adequate approximation to the true Nyquist plot. Generally, the pole approximating the single pole at the origin should be at least three orders of magnitude smaller than the magnitudes of the other poles. In this case the approximate Nyquist plot will be identical to the true one in the third and fourth quadrants.

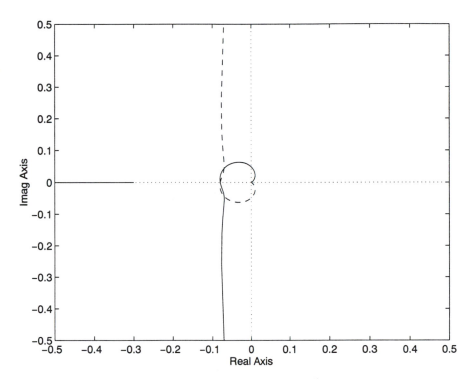

Figure 6.5 Close-up of the region around the origin in the Nyquist plot in Figure 6.4.

nyqnic.m

```
% nyqnic
% Creating custom scaled Nyquist and Nichols plots for a
% system with a pole at zero.
clf
[num,den]=zp2tf([-0.1+5*i; -0.1-5*i],[-0.001 -4 -0.2+10*i
-0.2-10*i], 100);
subplot(121), nyquist(num,den), axis([-20 20 -20 20])
subplot(122), nichols(num,den), axis([-360 0 -40 40])
ngrid
```

We have also included a pair of complex conjugate zeros that are close to the $j\omega$-axis and a lightly damped pair of poles. The automatically scaled Nyquist plot generated with the command **nyquist(num,den)** without any **axis** command is shown in Figure 6.6. We have not included the file that created this plot; however, the **num** and **den** are as shown in file *nyqnic.m*. Again, there are clues to the existence of a more interesting plot, but you must be observant to notice them. Customized plots, produced by run-

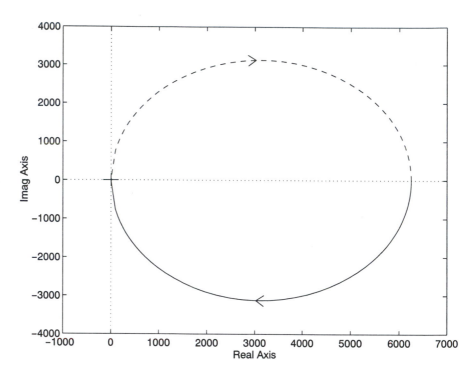

Figure 6.6 Automatically scaled Nyquist plot for $G(s) = 100(s + 0.1 - 5i)$ $(s + 0.1 + 5i)/(s + 0.001)(s + 4)(s + 0.2 - 10i)(s + 0.2 + 10i)$.

ning *nyqnic.m*, are shown in Figure 6.7. Notice, from the Nichols plot, that this system would have very little phase margin if it were in a unity gain feedback system. A precise estimate of the maximum gain for which the closed loop system is stable would require us to further expand the Nyquist plot in the vicinity of the origin. This is left as an exercise. Notice also that in the Nichols plot, the separate piece of the curve at the left really belongs at the right. MATLAB does not recognize that phase angles such as $-355°$ should be plotted as $+5°$ and not as $-355°$ in order to make the plot look better. The fact that phase angles repeat every $360°$ occasionally causes similar problems. You should watch for this.

We will give further examples of the need to customize the scaling of plots when we describe frequency-domain design methods. The remainder of this section deals with techniques for extracting data from frequency plots. All of the commands that generate frequency plots will return the data that create the plot instead of generating the plot if you enter the command with output arguments, for example, **[x,y,z]**, to the left of the equal sign. This is demonstrated in the file *getinfo.m*, where we compute the peak magnitude and bandwidth for a given system. In this

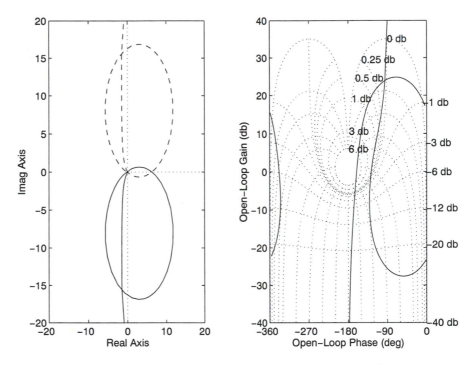

Figure 6.7 Blowup of region near the origin of the Nyquist plot from Figure 6.6 is shown at the left. Nichols plot for the same system is at the right.

file we have blacked out the values of **num** and **den** to give you a better sense of the way engineers often use Bode plots. If you know $G(s)$ exactly, that is, if you can write down **num** and **den**, you don't really need to extract exact peaks and bandwidth from the Bode plot. Extracting peaks and bandwidth from a Bode plot is more appropriate when you have only the Bode plot, for example, from experimental data, and not $G(s)$.

getinfo.m

```
% getinfo
% Getting information from Bode plots.
clf
num=■;   den=■;
bode(num,den)
[mag,phase,w]=bode(num,den);
[Mp,k]=max(mag)
wp=w(k)
n=1;
while 20*log10(mag(n))>=-3, n=n+1;end
bandwidth=w(n)
```

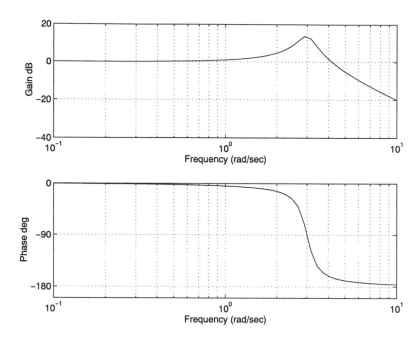

Figure 6.8   "Experimentally" determined Bode plot.

Notice that in file *getinfo.m* **bode** is entered twice. The first time it returns the plot in Figure 6.8. The second time it creates three vectors. The first vector, **mag**, contains the values of $|G(j\omega)|$. The second vector, **phase**, contains the values of $\angle\, G(j\omega)$. The third, and final, vector **w** contains the corresponding values of ω.

We next use **max** to find M_p, the peak magnitude of the frequency response. The variable **k** indicates the index number of **Mp**, that is, **Mp** = **mag(k)**. Note that we have made no attempt to write general-purpose code. We first created the Bode plot and then, knowing $|G(j\omega)|$ and the approximate values of M_p and the corresponding frequency, ω_p, we wrote two simple lines to compute them exactly. The last three lines of *getinfo.m* compute the bandwidth. Notice that **bode** returns the values of $|G(j\omega)|$ itself and not $20\log|G(j\omega)|$. Thus, in computing the bandwidth we had to use **20*log10(mag(n))**.

The result of running the file is

Mp =

    4.8966

k =

    22

```
wp =
     2.8994
bandwidth =
     4.7820
```

MATLAB contains one other command that specifically extracts data from Bode plots. This is the command **margin**. There are several ways to use **margin**. One way is to compute and store $|G(j\omega)|$, $\angle\ G(j\omega)$, and ω. This is done by entering **bode** with three left-hand arguments, say **[mag,phase,w]**. Then, entering **margin(mag,phase,w)** with no left-hand arguments produces a Bode plot with lines indicating gain and phase margins and the frequencies at which they occur. Entering the same command with four left-hand arguments as shown below returns the gain margin, phase margin, and the frequencies at which they are measured but does not produce a Bode plot:

>> **[gain_margin,phase_margin,wcg,wcp]=margin(mag,phase,w)**

You can also use **margin(num,den)** or **margin(a,b,c,d)** to create Bode plots and display the gain and phase margin for systems in transfer function or state-space form, respectively. It is interesting to note that if you run *get-info.m* and then enter **margin(mag,phase,w)** followed by **margin(num,den)** you get different answers. The difference is most significant for the gain margin. The answer obtained from **margin(mag,phase,w)** is wrong. That version of **margin** interpolates between points on the plot. The interpolation finds a finite frequency at which the phase angle crosses –180° even though no such finite frequency exists.

Finally, the command **ginput** usually allows you to use a mouse (or arrow keys) to select points on a graph and obtain their coordinates. However, there is a bug when you use this with Bode plots. The horizontal coordinate is returned correctly, but when you click on a point on the magnitude plot, the vertical one is not. Clicking on the phase plot works correctly.

There are more examples of data extraction from frequency-domain plots in the following section.

6.3 Design by Means of Frequency Response Plots

We begin by designing a simple controller for the system whose "experimentally determined" Bode plot is shown in Figure 6.9. Notice that this system has three more poles than zeros, as evidenced by the phase angle being asymptotic to –270° as $\omega \to \infty$. Since there is no peaking of the magnitude response, except at $\omega < 10^{-2}$, there are no lightly damped complex conjugate pairs of poles. You can clearly see that the system has a single pole at 0.001. You can also see that there are two more poles in the general region of $\omega = 1$ to $\omega = 30$. Experimentally determined Bode plots are rarely as clean as this. Noise and nonlinearity distort the plots, especially at higher frequencies. For this system assume the specifications the closed-loop system must satisfy are that the bandwidth be maximized subject to the constraint that the damping ratio be close to 0.707. Figure 6.9 was actually produced by the first three lines of *design1.m*.

We used **fbode** rather than **bode** to produce the plot because it uses a faster algorithm for diagonalizable systems. Our system has distinct poles, so it is diagonalizable.

design1.m
```
% design1
% A simple design example using the Bode plot.
clf
num=■;
den=■;
fbode(num,den)
[mag,phase,w]=bode(num,den);
n=1;
while phase(n)>=-115; n=n+1; end
gain=mag(n)
gain_to_add=1/gain
```

We have again blacked out the values of **num** and **den** to enhance the impression that the Bode plot is experimental data and that a precise analytical model of $G(s)$ is not available (i.e., given experimental data we would enter the vectors **mag**, **phase**, and **w** directly from the data). For a second-order system we know that a 0.707 damping ratio corresponds to a phase margin of 65.52°. Even though our system is not second order, it is reasonable to try a controller that consists of a pure gain and to choose

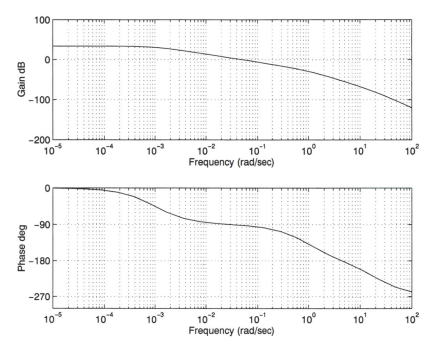

Figure 6.9   "Experimentally" determined Bode plot.

that gain so that the system has a phase margin close to 65°. The last five lines of *design1.m* calculate the required gain, which is 13.7815. In the real world this would probably mean a gain of 14.

We can test this design by computing the Bode plot of the closed-loop transfer function. The essential computation is the following:

$$G_{cl}(j\omega) = \frac{kG(j\omega)}{1 + kG(j\omega)}$$

where

$$G(j\omega) = \text{the open-loop transfer function}$$

$$G_{cl}(j\omega) = \text{the closed-loop transfer function}$$

$$k = 13.7815 \text{ (as computed earlier)}$$

for each of the frequencies ω for which data are available. $G(j\omega)$ is calculated from the magnitude $|G(j\omega)|$ (**mag**) and the phase angle $\angle G(j\omega)$ (**phase**) as

$$G(j\omega) = |G(j\omega)|e^{j(\angle G(j\omega))}$$

A MATLAB program to do this is given in file *closedloop.m*. The file assumes that **mag, phase**, and **w** are already available, as they are in our case because file *design1.m* was executed just before running *closedloop.m*.

closedloop.m

```
% closedloop
% A calculation and display of the Bode plot of the
% closed-loop system resulting from applying design1.
clf
phaserads=phase*2*pi/360;
gcl=13.7815*mag.*exp(i*phaserads)./ ...
(1+13.7815*mag.*exp(i*phaserads));
magcl=abs(gcl);
temp_phase=angle(gcl);
phasecl=unwrap(temp_phase)*360/(2*pi);
subplot(211)
semilogx(w,20*log10(magcl))
grid
title('Bode magnitude plot of closed-loop system'),
xlabel('Frequency (rad/s)'), ylabel('Gain (dB)')
subplot(212)
semilogx(w,phasecl)
grid
title('Bode phase plot of closed-loop system'),
xlabel('Frequency (rad/s)'),ylabel('Angle (Degrees)')
```

There are several tricky aspects to computing the closed-loop transfer function. First, note that **phase** is in degrees but the exponential command **exp** requires that its argument be in radians. The second line of the file converts **phase** in degrees to **phaserads** in radians. The third line calculates the closed-loop transfer function $G_{cl}(j\omega)$ as already described and stores it in the variable **gcl**. The fourth line generates $|G_{cl}(j\omega)|$ as **magcl**, and the fifth line uses the command **angle** to compute the closed-loop phase angle. Note that **angle** returns the phase angle in radians.

Furthermore, **angle** gives all the angles it computes values between $-\pi$ and π radians. This is mathematically correct but different from the usual rule for computing phase angles. The sixth line uses the command **unwrap** to convert the angle to negative radians (i.e., between 0 and -2π) and then converts the angle to degrees. The rest of the file produces the Bode plots using plotting techniques described earlier.

Running the file produces Figure 6.10. The fact that there is no peaking in the closed-loop magnitude plot indicates that we have come close to our design goal. To further demonstrate that we have a good design we abandon the pretense that we do not have an analytic expression for $G(s)$

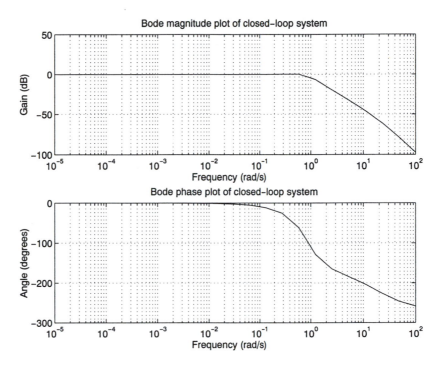

Figure 6.10   Bode plot of the closed-loop system designed in file *design1.m.*

and enter the following three lines, which clear the figure and plot the closed-loop step response.

```
>>clf
>>[numcl,dencl]=cloop(13.7815*num,den);
>>step(numcl,dencl)
```

We assume that **num** and **den** are already available from having run *design1.m.* The closed-loop step response is shown in Figure 6.11. Its behavior is quite close to the required damping. Would it be possible to increase the bandwidth significantly while maintaining a step response with a 0.707 damping ratio?

There are several approaches that could be used to answer this question. One way is illustrated with file *design1nic.m.* The file shows you the $G(s)$ that we used in *design1.m.* It is

$$G(s) = \frac{1}{s(s+1)(s+20)}$$

with a slight left shift of the pole at $s = 0$ to simplify computations. It is easy, and a worthwhile exercise, to verify both theoretically and by ex-

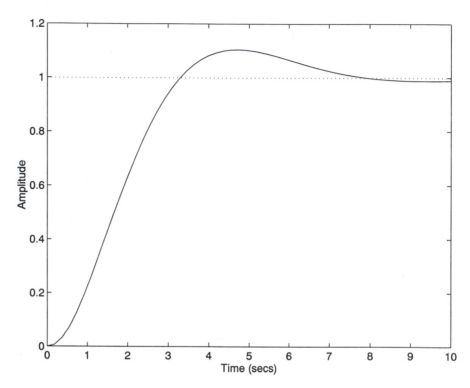

Figure 6.11   Closed-loop step response of the system shown in Figure 6.9 when feedback gain of 13.7815 is used.

perimentation using MATLAB that the design is independent of the left shift provided the left shift is smaller than 0.001. The file *design1nic.m* produces Figure 6.12, which shows two Nichols plots. The solid line displays the same frequency response as Figure 6.9. The dashed line shows the same frequency response except that the gain is 100 times larger. Note the way *design1nic.m* customizes the Nichols plot. The **axis** command is used to provide the appropriate ranges. Then **nichols** is used only to calculate the response data. The plots are actually produced with the **plot** command. This allows us to specify two different line textures for the two different lines. Notice that the **plot** command has six input arguments. The first three arguments define the x-coordinate vector, the y-coordinate vector, and the texture, respectively, for the first line to be plotted. The last three arguments specify the x-coordinate vector, the y-coordinate vector, and the texture, respectively, for the second line to be plotted. We could have used the **hold** command to superimpose the two curves more simply, but both curves would have been identical in color and texture.

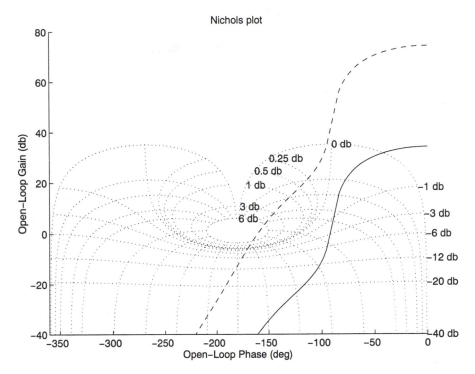

Figure 6.12   Nichols plot of the system shown in Figure 6.9 (solid curve) and the same system with a gain of 100 (dashed curve).

The **hold on** makes the previous **axis** command apply until the **hold off** at the end of the script. Alternatively, you could put the **axis** command at the very end, after **ngrid**, and delete the **hold on** and **hold off** commands. It is necessary to include the lines **phase(1)=0;** and **phase1(1)=0;** because MATLAB assigns them values of –360°. This is correct but (1) it is a rather odd way to write 0° and (2) it creates two unsightly lines on the resulting plot.

design1nic.m

```
% design1nic
% Nichols plots of 1/s(s+1)(s+20) for gains of 1 and 100.
clf
num=1;
den=conv([1 20],conv([1 0.001], [1 1]));
axis([-360 0 -40 80])
hold on
[mag,phase,w]=nichols(num,den);
[mag1,phase1,w1]=nichols(100*num,den);
```

```
phase(1)=0;
phase1(1)=0;
plot(phase,20*log10(mag),'-',phase1,20*log10(mag1),'- -')
title('Nichols plot')
ngrid
hold off
```

The dashed curve in Figure 6.12 shows that a gain of 100 would be much too large to maintain the required damping. The curve cuts the 6-dB ellipse on the Nichols chart, indicating substantial peaking of the closed-loop frequency response. This indicates an underdamped step response. A gain of 10 would produce a curve midway between the two displayed in Figure 6.12. Remember that a pure gain translates the Nichols plot vertically. The curve corresponding to a gain of 10 would be nearly tangent to the 0-dB curve on the Nichols chart, indicating that a gain of 10 would provide about the best performance. Thus, our original design that uses a gain of about 14 is close to the best possible. Finally, if you examine *design1nic.m* you will notice that Figure 6.12 could have been produced directly from **mag, phase,** and **w** [i.e., without knowing $G(s)$].

Our second example repeats the controller design for the third-order system considered in Chapter 5 as a design example. The design begins with the generation of the Bode plot for the given system. This is done, using a gain of 16, in file *leadlagexstep1.m*, that is,

$$G(s) = \frac{16}{s(s + 1)(0.2s + 1)}$$

leadlagexstep1.m
```
% leadlagexstep1
% This generates the Bode plot for 16/s(s+1)(0.2s+1).
clf
num=16;
den=conv([1 0],conv([1 1], [0.2 1]));
bode(num,den)
```

We base the design on the following specifications:

1. The open-loop low-frequency gain should be as large as possible.

2. The closed-loop damping ratio should be 0.707.

3. The settling time of the closed-loop system should be 3 seconds.

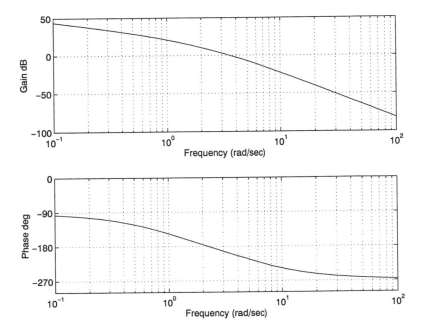

Figure 6.13   Bode plot of $G(s) = 16/s(s+1)(0.2s+1)$.

These specifications are somewhat artificial. One would normally be given a lower bound on the open-loop gain, an upper bound on the settling time, and a range of acceptable damping ratios. Our choice of specifications simplifies the design process and makes it easier to explain our design steps.

One problem in describing the design process in a textbook is that there is a considerable amount of trial and error in really doing a design. It is impractical to include several false starts toward the final design in a textbook. Thus, textbook design examples are almost always incredibly efficient. Success is achieved on the first try. In order to do that, we need to cheat. Since we already have a reasonably successful controller design for the third-order system considered here, we know that a gain of 16 combined with a lead-lag compensator can probably satisfy our specifications and that a higher gain is probably not feasible. That is why we used a gain of 16 in the Bode plot of Figure 6.13. In a real design this would be an initial guess and we might well need several tries before we arrived at a reasonable choice of gain.

The next step is to design the lead-lag compensator. In Chapter 5 we assumed the lead and lag could be designed independently. This is possible if amplifiers are used to isolate the lead compensator electronically from the lag compensator. Sometimes it is undesirable to include such amplifiers. In applications requiring low noise, for example, it is preferable to use only passive components (resistors, capacitors, and inductors) because they produce less noise than amplifiers. It is possible to build lead-lag compensators using only passive components. The price to be paid is that such compensators allow a limited difference between the lead and lag break points in comparison to active (amplifier) lead-lag compensators. Assume that we want to design a passive lead-lag compensator. The next step is to determine the closed-loop bandwidth needed to satisfy the damping ratio and settling time specifications. A common approach for doing this is to assume the closed-loop system can be approximated by a second-order system. The bandwidth of a second-order system having a given settling time and damping ratio can be computed from the following formula (see Nise, Chapter 10 for a detailed discussion).

$$\omega_{BW} = \frac{4}{T_s \zeta} \sqrt{(1 - 2\zeta^2) + \sqrt{4\zeta^4 - 4\zeta^2 + 2}}$$

where

$$\zeta = \text{damping ratio}$$
$$T_s = \text{settling time}$$
$$\omega_{BW} = \text{bandwidth}$$

This formula is implemented in file *leadlagexstep2.m*. The result is shown immediately below the file.

leadlagexstep2.m

```
% leadlagexstep2
% This defines the specifications and computes
% the bandwidth for our current design.
zetaspec=0.707;
settling_time_spec=3;
omegabw=4*sqrt(1-2*zetaspec^2 ...
+sqrt(4*zetaspec^4-4*zetaspec^2+2))/ ...
(zetaspec*settling_time_spec)

omegabw =
      1.8862
```

The next step is to decide on the parameter values of the compensator. The transfer function of a passive lead-lag compensator $G_c(s)$ is given by (see Nise, Chapter 11, for further details):

$$G_c(s) = \left(\frac{s + \frac{1}{T_{le}}}{s + \frac{\gamma}{T_{le}}}\right)\left(\frac{s + \frac{1}{T_{la}}}{s + \frac{1}{\gamma T_{la}}}\right)$$

where $\gamma > 1$, lead time constant T_{le}, and lag time constant T_{la} are the parameters to be determined.

Looking at Figure 6.13, we see that the gain crossover frequency is at about $\omega = 3.3$ rad/s. Since we need a bandwidth of just under 2 rad/s, we can afford a small decrease in gain near the gain crossover frequency [i.e., the value of ω at which $20\log|G(j\omega)|$ passes through 0 dB]. Such a decrease in gain would lower the gain crossover frequency to approximately $\omega = 2$. Since a damping ratio of 0.707 requires a phase margin of approximately $65°$, we see that we need to reduce the phase shift in the vicinity of $\omega = 2$ rad/s by about $65°$. This is a lot of phase shift. Based on the results reported in Nise, Chapter 11, such a phase shift would be nearly achievable with $\gamma = 30$ but, for reasons we will explain shortly, we believe $\gamma = 10$ is preferable. Looking at file *leadlagexstep3.m* you will see that we chose **gamma** $= \gamma = 10$.

The lag time constant T_{la} is not very critical to the design. The lag compensator is basically there to keep the low-frequency gain high without causing instability. It is somewhat easier to think of the lag time constant in terms of its corresponding frequency $\omega_{la} = 2\pi/T_{la}$. Typically, ω_{la} is chosen to be about $0.1(\omega_{BW})$. We have chosen **Tla** $= T_{la} = 2\pi/0.1(\omega_{BW})$, as can be seen in file *leadlagexstep3.m*.

Finally, a value for T_{le} is determined as follows. The frequency at which we want the largest positive phase angle from the lead-lag compensator, ω_{max}, is chosen to be ω_{BW}. Once this frequency is known, T_{le} **(Tle)** is given by the following formula (from Nise, Eq. 11.9).

$$T_{le} = \frac{1}{\omega_{max}\sqrt{\beta}}$$

where

$$\beta = 1/\gamma$$

The complete compensator transfer function $G_c(s)$ described by **numc** and **denc** is computed using the **series** command to connect the lead and lag components of the compensator, which themselves have been determined using the **zp2tf** command. We then plot the Bode plot for the compensator alone, as shown in Figure 6.14. Note that the peak positive phase contribution from the compensator is at $\omega \approx 2$ rad/s as desired. Note that the title appears in the wrong place, over the phase plot, not at the top of the figure. You could fix this by inserting the line **subplot(211)** just before the last line of *leadlagexstep3.m*.

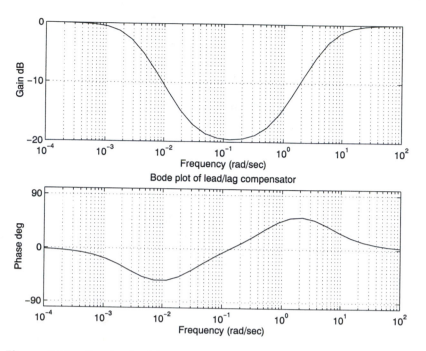

Figure 6.14   Passive lead-lag compensator.

leadlagexstep3.m

```
% leadlagexstep3
% Design of lead-lag compensators.
clf
% lag compensator design
Tla=2*pi/(0.1*omegabw);
beta=0.1;   gamma=1/beta;
[numla,denla]=zp2tf(-1/Tla,-1/(gamma*Tla),1);
% lead compensator design
omegamax=omegabw;
Tle=1/(omegamax*sqrt(beta));
[numle,denle]=zp2tf(-1/Tle,-gamma/Tle,1);
[numc,denc]=series(numle,denle,numla,denla);
bode(numc,denc)
title('Bode plot of lead-lag compensator')
```

We next display in Figure 6.15 the Bode plot of the compensated system, $G_c(s)G(s)$, generated by file *leadlagexstep4.m*. Note again that the **series** command is used to connect the compensator $G_c(s)$ and the open-loop system $G(s)$.

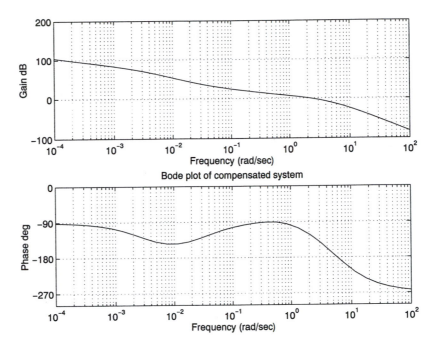

Figure 6.15 Bode plot of system plus lead-lag compensator.

leadlagexstep4.m
```
% leadlagexstep4
% Bode plot of compensated system.
clf
[numt,dent]=series(numc,denc,num,den);
bode(numt,dent)
title('Bode plot of compensated system')
[mag,phase,w]=bode(numt,dent);
```

Finally, we check the performance of our design by computing and displaying the step response of the closed-loop system. This is done in file *leadlagexstep5.m*, and the result is shown in Figure 6.16. Note the way we expanded the relevant portion of the plot using Handle Graphics. The **axis** command would have worked equally well.

leadlagexstep5.m
```
% leadlagexstep5
% Plots the step response of the closed-loop system
% we designed.
clf
```

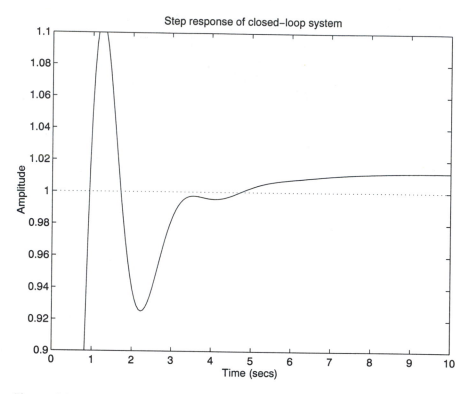

Figure 6.16   The step response of the closed-loop system is then used to check the design.

```
[numcl,dencl]=cloop(numt,dent);
step(numcl,dencl)
% The next 4 commands use Handle Graphics to expand the
% relevant region of the plot.
h=gcf;
h1=get(h,'CurrentAxes');
set(h1,'Ylim',[.9 1.1])
set(h1,'Xlim',[0 10])
title('Step response of closed-loop system')
```

Notice, in Figure 6.16, that the settling time specification is met exactly. However, the response is clearly underdamped. We have failed to meet the requirement of 0.707 damping ratio. We could choose y=30 to try to improve the damping ratio. However, this might not be a good idea. Look at Figure 6.15 again. At $\omega \approx 7 \times 10^{-3}$ the phase angle is below –140°. Increasing y will make this phase angle even more negative. If the phase angle becomes more negative than –180° we will have instability. The phase angle of a physical implementation of this system might well have a few degrees of additional phase shift. Furthermore, increasing y will move the gain crossover frequency [the value of ω at which $20 \log|G(j\omega)|$ passes through 0 dB] to the left, slowing the closed-loop response.

We leave the job of improving our design as an exercise. All in all, it is not easy. You could try placing the peak of the compensator phase at a slightly higher frequency. You could also try an active lead-lag compensator. Large improvement may require a more elaborate compensator such as those described in the next chapter.

6.4 Exercises

1. Create a Nyquist plot for the system $G(s) = 1/(s+20)(s+0.001)(s+1)$. Blow up the region near the origin. Display just the region $-0.003 \le \text{Re } G(j\omega) \le 0$ and $-0.0005 \le \text{Im } G(j\omega) \le 0.0005$. How did we know this region would be interesting?

2. Draw Bode, Nyquist, and Nichols plots for a sixth-order Pade approximation to a pure delay of 1 second.

3. Recreate Figures 6.4 and 6.5 using $s = -0.1, -0.01, -0.005, -0.0009$ to approximate the pole at $s = 0$. Create Bode plots for the same $G(s)$ using these approximations to the pole at $s = 0$. Use the theoretical asymptotes for first- and second-order factors in Bode plots to explain your results.

4. Try to improve on the controller we designed for the system $G(s) = 1/s(s + 1)(0.2s + 1)$. In particular, try increasing **omegabw** slightly; say to 3.

5. One way to improve a controller is to buy a better sensor and/or actuator. Assuming the actuator in Exercise 3 is a motor and that the pole at -1 is due to a potentiometer with a lot of friction, repeat Exercise 3 with poles at 0, -3, -5. This is meant to approximate the effect of buying a better potentiometer.

6. Repeat Exercise 4 with $G(s) = 1/s(s + 1)(0.1s + 1)$. This is meant to approximate the effect of a better motor.

7. Construct an example of an open-loop $G(s)$ for which gain and phase margins are large but the closed-loop system is nearly unstable. (Hint: See Figure 6.15.) Produce Bode, Nyquist, and Nichols plots of this system's transfer function.

8. Repeat Exercise 5.5, the design of a control system including a notch filter, in the frequency domain. Create Bode, Nyquist, and Nichols plots for the system plus the notch filter. Create a Bode plot of the notch filter alone to see how the filter got its name.

9. Create some interesting looking Nyquist and Nichols plots. One such example might be the plots for

$$G(s) = \frac{100(s + 1)^2}{(s + 0.1)(s + 10)(s^2 + 1.65 + 400)}$$

10. Create Bode, Nyquist, and Nichols plots for the linear transfer function in *constabex.m*, see Figure 4.10. What can you say about the frequency response plots for conditionally stable systems?

7 STATE-SPACE COMPUTATIONS

State-space computations provide a powerful alternative to transfer function methods for the analysis and design of control systems. The state-space approach has been particularly useful in problems where

1. Good mathematical models of the system to be controlled are available.

2. The state dimension is not too high (the exact number depends on many factors and is continually increasing but a maximum of approximately ten states is a reasonable rule of thumb today).

3. The controlled system has to satisfy stringent specifications.

When these conditions are satisfied the state-space approach facilitates design of high-performance closed-loop systems at the cost of a more complicated controller. Aerospace problems provide a number of examples.

MATLAB provides many commands for state-space computations that not only allow you to construct controllers in state-space form but also help you to select control parameters so that your system will meet performance requirements. This chapter shows you how to use these commands to investigate and design control systems in the state space.

7.1 Controllers and Observers

Consider the third-order plant described by the transfer function $G(s) = 1/s(s+1)(0.2s+1)$. In Chapter 5, Section 5.3, we created root locus plots to design a controller for this plant so that the closed-loop system would have a damping ratio of 0.707 with the largest possible natural frequency. Since we were working with transfer functions, our design efforts were limited to adding poles and zeros to the system by means of a compensator so that the resulting closed-loop system poles would achieve the desired performance. With a lead-lag compensator we were able to produce a system with a 0.707 damping coefficient and a natural frequency of approximately 2.6.

Using a state-space description of the plant, on the other hand, we can design a controller and an observer to place the closed-loop poles as desired. We demonstrate doing so in the next example with file *regulator.m*, and we introduce the commands **acker** and **place** for controller and observer design. To begin with, we transform $G(s)$ into state-space form. The first three lines of file *regulator.m* convert $G(s)$ into a (controller canonical) state-space model.

regulator.m

```
% regulator
% Example of state-space controller and observer design.
% First enter the transfer function G(s).
numG=1;
denG=conv(conv([1 0], [1 1]),[0.2 1]);
% Convert to state space.
[Ag,Bg,Cg,Dg]=tf2ss(numG,denG);
% Generate a second-order system.
damping=0.707;
wn=3;
[num2,den2]=ord2(wn,damping);
% Select desired poles to include poles of second-order
% system.
dominant=roots(den2);
desiredpoles=[dominant' 10*real(dominant(1))];
% Compute the controller gain K.
K=acker(Ag,Bg,desiredpoles);
% Compute the closed-loop state-feedback system.
Asf=Ag-Bg*K; Bsf=Bg; Csf=Cg; Dsf=0;
[numsf,densf]=ss2tf(Asf,Bsf,Csf,Dsf);
% Select observer poles to be 10 times faster than
% controller.
observerpoles=10*desiredpoles;
```

```
% Compute observer gain L.
L=acker(Ag',Cg',observerpoles);
% Compute the closed-loop system with controller and
% observer.
Areg=[(Ag-Bg*K) Bg*K; zeros(size(Ag)) (Ag-L'*Cg)];
Breg=[Bg; zeros(size(Bg))];
Creg=[Cg zeros(size(Cg))];
Dreg=0;
[numreg,denreg]=ss2tf(Areg,Breg,Creg,Dreg);
damp(denreg);
```

To see this state-space model of the original system, after running *regulator.m*, type

>>**printsys(Ag,Bg,Cg,Dg)**

The output will look as follows:

```
    a =
                            x1         x2        x3
                    x1    -6.00000  -5.00000       0
                    x2     1.00000         0       0
                    x3           0   1.00000       0
    b =
                            u1
                    x1     1.00000
                    x2           0
                    x3           0
    c =
                            x1         x2        x3
                    y1           0         0   5.00000
    d =
                            u1
                    y1           0
```

You can verify that this model corresponds to $G(s)$ by checking that the eigenvalues of the matrix **Ag** correspond to the poles of $G(s)$ as follows:

```
>>eig(Ag)
ans =
             0
        -1.0000
        -5.0000
```

Alternatively, type

```
>> 5*denG
ans =
      1    6    5    0
```

to find that the coefficients of the original system's characteristic equation $s^3 + a_2 s^2 + a_1 s + a_0 = 0$ are $a_2 = 6$, $a_1 = 5$, and $a_0 = 0$. Then by inspection you can see that the first row of **Ag** corresponds to $[-a_2 - a_1 - a_0]$ as expected for the controller canonical state-space model of $G(s)$.

Now suppose that we want to improve upon the system performance with the lead-lag compensator of Section 5.3 by designing a system with a closed-loop damping coefficient of 0.707 and a natural frequency, ω_n, of 3.0. The **ord2** command in the file *regulator.m* generates a second-order system with the desired dynamic characteristics. For our desired closed-loop third-order system we choose the poles of this desired second-order system to be the two dominant poles, and we select the third pole to be faster than the dominant poles by about a factor of 10. The vector **desiredpoles** is created with elements equal to these three desired pole locations,

```
>> desiredpoles
desiredpoles =
    -2.1210 - 2.1216i    -2.1210 + 2.1216i    -21.2100
```

If we first assume that the states in our system are available to the controller, we can design a state-feedback controller K as shown in Figure 7.1 to generate a system with the desired closed-loop poles.

The MATLAB command **acker(Ag,Bg,desiredpoles)** computes the constant feedback gain vector K for the state-space system described by **Ag** and **Bg** and the closed-loop pole locations of the vector **desiredpoles**. For our example K is calculated to be

```
K =
    19.4520    93.9728    190.8900
```

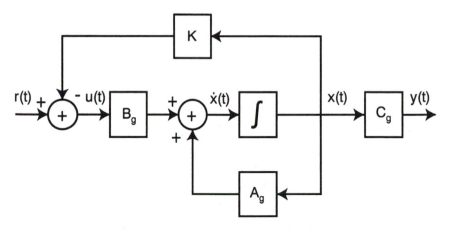

Figure 7.1   State-feedback control system.

The command **place** can be used instead of **acker**. In fact, **place** is rec-
ommended in the MATLAB manual rather than **acker** because it uses a
more reliable algorithm. However, **acker** is reliable for systems of order
less than six. Additionally, **place** has trouble placing poles that have multiplicity greater than the number of inputs. As a result, we use **acker**
throughout this chapter. The closed-loop system of Figure 7.1 can be
described by

$$\dot{x} = (A_g - B_g K)x + B_g r = A_{sf}x + B_{sf}r$$
$$y = C_g x = C_{sf}x$$

(since $D_g = 0$). We calculate this closed-loop state-feedback model for our
system (**Asf, Bsf, Csf, Dsf**) as well as the closed-loop transfer function
description (**numsf,densf**) in *regulator.m*. To confirm that the poles of
the closed-loop system are as desired, enter

```
>> roots(densf)
ans =
  -21.2100
    -2.1210 + 2.1216i
    -2.1210 - 2.1216i
```

Since these poles match the desired poles, the system described by **numsf**
and **densf** will achieve the desired damping and natural frequency.

Suppose now that the states are not available to the controller. In this case
we need to design an observer to estimate the states x from the output y.

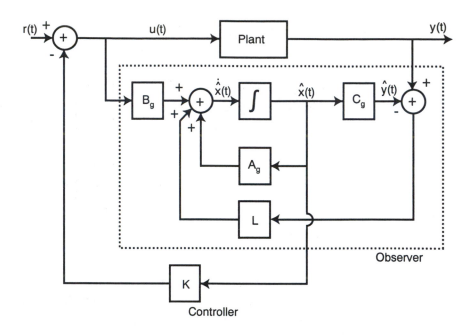

Figure 7.2   System with controller and observer.

We will feed back these estimated states $\hat{x}$ to the controller K as shown in Figure 7.2

Since we would like the estimated states to converge to the actual states faster than the transient response of the plant, we select the observer poles to be ten times faster than the controller poles. The elements of the vector **observerpoles** in the file *regulator.m* correspond to these desired observer poles.

To compute the constant observer gain vector L, we again use the command **acker** (alternatively, **place**).   This time, however, we utilize **L=acker(Ag',Cg',observerpoles)**. The computed vector **L** is

L =

     1.0e+04 *

     2.7848   0.1680   0.0050

The equations for the closed-loop system of Figure 7.2 can then be written as

$$\begin{pmatrix} \dot{x} \\ \dot{e} \end{pmatrix} = \begin{bmatrix} (A_g - B_g K) & B_g K \\ 0 & (A_g - LC_g) \end{bmatrix} \begin{pmatrix} x \\ e \end{pmatrix} + \begin{bmatrix} B_g \\ 0 \end{bmatrix} r = A_{\text{reg}} \begin{pmatrix} x \\ e \end{pmatrix} + B_{\text{reg}} r$$

$$y = [C_g \ 0] \begin{pmatrix} x \\ e \end{pmatrix} = C_{\text{reg}} \begin{pmatrix} x \\ e \end{pmatrix}$$

where $e = x - \hat{x}$ and $D_{\text{reg}} = 0$. This closed-loop state-space model is computed at the end of *regulator.m* as (**Areg, Breg, Creg, Dreg**) and converted into a transfer function description (**numreg,denreg**). The output of *regulator.m* shows the damping characteristics of the closed-loop system using the command **damp**.

```
>> regulator
```

Eigenvalue	Damping	Freq. (rad/sec)
1.0e+02 *		
-0.0212 + 0.0212i	0.0071	0.0300
-0.0212 - 0.0212i	0.0071	0.0300
-0.2121 + 0.2122i	0.0071	0.3000
-0.2121 - 0.2122i	0.0071	0.3000
-0.2121	0.0100	0.2121
-2.1210	0.0100	2.1210

As you can see, this transfer function provides the desired performance characteristics, that is, the first two eigenvalues correspond to the dominant poles and have a damping coefficient of 0.707 and natural frequency of 3.0. The remaining poles are considerably faster.

Figure 7.3 shows a unit step response for the closed-loop state-space system as compared to a unit step response for the closed-loop system with a lead-lag compensator designed using the root locus method in Chapter 5. The plot is generated by running file *compare.m*.

compare.m
```
% compare
% Comparison plots of root locus lead-lag and
% state-space designs.
clf
numoll=16.2*conv([1 1], [20 1]);
denoll=conv(denG,conv([0.1 1], [200 1]));
[numcll,dencll]=cloop(numoll,denoll);
[y,x,t]=step(numcll,dencll);
plot(t,y,'c- -')
```

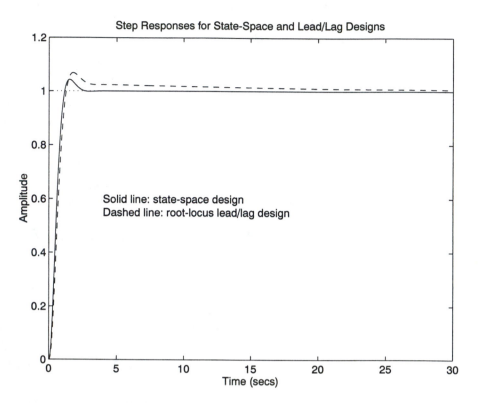

Figure 7.3   Step responses for state-space and lead-lag designs.

```
hold on
step(Areg,Breg,Creg/dcgain(numreg,denreg),Dreg,1,t);
text(4,0.6,'Solid line:   state-space design')
text(4,0.55,'Dashed line:   root-locus lead-lag design')
hold off
```

As planned, our system with controller and observer responds faster, since the natural frequency (3.0 rad/s) is greater than that of the system with the lead-lag compensator (2.6 rad/s). The damping coefficient is the same for both systems. Note that, in the eighth line after the comments of *compare.m*, we have added a gain to the closed-loop state-space system to eliminate the steady-state error. A better design would use integral control.

7.2 Controllability and Observability

The fact that we started with a transfer function description in the previous example guaranteed that we could find a minimal state-space realization for our plant, that is, one that was both controllable and observable. In the case that a derived state-space model included redundant states corresponding to poles and zeros that would cancel each other out in the transfer function, these states could be removed with the MATLAB command **minreal**. For example, suppose the plant was described as $G_r(s) = (s+3)/s(s+1)(0.2s+1)(s+3)$ which is identical to the previous example except for an extra pole and zero at –3. Then the corresponding state-space model **Agr**, **Bgr**, **Cgr**, and **Dgr** created by running the file *minimize.m* would have a redundant state.

minimize.m

```
% minimize
% Example of a state-space model with an unobservable
% or uncontrollable state and how to remove it.
numgr=[1 3];
dengr=conv(conv(conv([1 0], [1 1]), [0.2 1]), [1 3]);
[Agr,Bgr,Cgr,Dgr]=tf2ss(numgr,dengr);
[Amin,Bmin,Cmin,Dmin]=minreal(Agr,Bgr,Cgr,Dgr);
```

We can illustrate this by checking the order of the system (after running *minimize.m*):

```
>>length(eig(Agr))
ans =
     4
```

which is greater by 1 than in the previous example. The last line in *minimize.m* shows how to use the **minreal** command to produce the state-space model without the extra state. Verify that the order of the minimized system is the same as for the plant in Section 7.1.

If instead of starting with a transfer function we start with a state-space model of a given system, it is possible that our model will be neither controllable nor observable. If this is the case, it may not be possible to place as desired the poles corresponding to the uncontrollable modes or to estimate the states corresponding to the unobservable modes.

MATLAB provides several commands that allow you to determine whether or not a state-space model is controllable and/or observable. These include **ctrb**, **obsv**, **gram**, and **rank**, which are described in the following example. Consider the second-order plant described by the state-space model

$$A = \begin{bmatrix} -5 & -2 \\ -3 & 0 \end{bmatrix}, \qquad B = \begin{bmatrix} 1 \\ -3 \end{bmatrix}, \qquad C = [1 \quad -1], \qquad D = 0$$

Note that this system is unstable because A has a positive eigenvalue:

```
>>eig([-5 -2; -3 0])
ans =
      -6
       1
```

To check for controllability we need to examine the rank of the controllability matrix $C_M = [B \; AB]$. The MATLAB command **ctrb(A,B)** computes the controllability matrix corresponding to the system described by (**A,B**). In *controllable.m* we perform this computation and then compare the rank of C_M to the order of the system.

controllable.m
```
% controllable
% Demonstration of the test for controllability.
A=[-5 -2; -3 0];
B=[1; -3];
rank_of_Cm=rank(ctrb(A,B))
model_order=length(A)
```

If you enter **controllable**, MATLAB will respond with

```
rank_of_Cm =
       1
model_order =
       2
```

As you can see, the system is uncontrollable since rank(C_M) < 2.

We can similarly check to see if the system is observable using the command **obsv(A,C)**, which computes the observability matrix $O_M = [C' \; A'C']'$. In *observable.m* we perform this computation and then check the rank of O_M.

observable.m

```
% observable
% Demonstration of the test for observability.
A=[-5 -2; -3 0];
C=[1 -1];
rank_of_Om=rank(obsv(A,C))
```

If you enter **observable**, MATLAB will respond with

rank_of_Om =
    2

Since rank$(O_M) = 2$, the system is observable.

An alternative way to check for controllability involves performing a coordinate change to bring the given state-space model into diagonal or modal form and then making the determination by inspection. Since controllability is unaffected by coordinate changes, the system will be controllable if and only if the modal state-space model is controllable. To transform our system into modal form we use the MATLAB command **canon** as shown in the file *modalform.m*.

modalform.m

```
% modalform
% Demonstration of the modal form.
A=[-5 -2; -3 0];
B=[1; -3];
C=[1 -1];.
D=0
[Am,Bm,Cm,Dm]=canon(A,B,C,D,'modal');
```

After running *modalform.m*, we inspect the matrix **Am** and the vector **Bm**.

```
>> Am
Am =
      -6     0
       0     1
>> Bm
Bm =
       0
       3.1623
```

As you can see, **Am** is a diagonal matrix since the eigenvalues of the system are distinct. Therefore, since the vector **Bm** has a zero element, one of the states of the system is not controllable. This confirms our previous determination of the uncontrollability of our system.

Inspection of the diagonal form of the state-space model for controllability will be straightforward only if the system has distinct eigenvalues, that is, no repeated eigenvalues. As a result, computing and checking the rank of the controllability matrix is generally a more reliable method. However, the matrices produced by **ctrb**, as well as **obsv**, tend to be ill-conditioned. Using **gram** to compute the controllability and observability grammians is a numerically more reliable way to determine controllability and observability. By entering

```
>>rank(gram(A,B))
ans =
      1
```

we see that the rank of the controllability grammian is less than 2, therefore, as we already know, the system is uncontrollable. Similarly,

```
>> rank(gram(A',C'))
ans =
      2
```

shows us that the rank of the observability grammian is 2, which indicates that the system is observable.

Since our state-space system is not controllable, we cannot design a controller to place the two system poles as we desire. However, because the rank of C_M is equal to one, we know that we can control one of the two states. If the uncontrollable state is stable, then a controller that affects only the controllable state would at least be able to stabilize the system. The MATLAB command **ctrbf(A, B, C)** finds a similarity transformation T of the given state-space model such that the transformed system is of the form

$$A_t = TAT' = \begin{bmatrix} A_{uc} & 0 \\ A_{21} & A_c \end{bmatrix}, \qquad B_t = TB = \begin{bmatrix} 0 \\ B_c \end{bmatrix}, \qquad C_t = CT'$$

where $(A_{uc}, 0)$ is uncontrollable, (A_c, B_c) is controllable, and $C(sI-A)^{-1}B = C_c(sI-A_c)^{-1}B_c$. To perform the transformation for our system (assuming **A, B, C,** and **D** have been generated, e.g., by running *modal.m*), enter

```
>>[At,Bt,Ct,T,k] = ctrbf(A, B, C)
```

MATLAB will respond with

```
At =
      -6.0000     -0.0000
       1.0000      1.0000
Bt =
      -0.0000
       3.1623
Ct =
       0.6325      1.2649
T =
       0.9487      0.3162
       0.3162     -0.9487
k =
       1     0
```

where **k** is a vector with elements equal to the number of independent column vectors in each of the blocks of the controllability matrix [Bt At*Bt]. That is, **Bt** has rank 1, so **k(1)**= 1. **At*Bt=Bt**, so **At*Bt** is dependent on **Bt**; therefore, **k(2)**=0. The sum of the elements of **k** is the rank of the controllability matrix, which in this case is 1.

As you can see from the matrix **At**, A_{uc} = -6. Since A_{uc} is negative we know that the uncontrollable mode of our system is stable. Similarly, by inspection we find that A_c = 1, which is positive and, therefore, unstable. Since this unstable mode is controllable, we can design a controller to place the corresponding closed-loop pole on the left-hand side of the complex plane to make the closed-loop system stable. That is, although we cannot place both the poles as we please, we can design a controller to stabilize the given system.

The MATLAB command **canon** can be used as before to generate the modal state-space model for a given system. In addition, **canon** can be used to transform a controllable system to the companion canonical form. For example, to transform the system of Section 7.1 to companion canonical form, type **[Ac,Bc,Cc,Dc]=canon(Ag,Bg,Cg,Dg,'companion')**. The coefficients of the characteristic polynomial of the system will comprise the right column of **Ac**. Unfortunately, the resulting companion form is poorly conditioned because MATLAB uses the ill-conditioned controllability matrix C_M to perform the transformation. As a result, you need to be very careful when using **canon** to produce the companion canonical state-space model.

The controllability and observability grammians can also be used for other purposes besides checking controllability and observability. For instance, the grammians are useful for identifying fast states in state-space models. Such states can often be neglected, thereby reducing model order. The file *reduce.m* uses **balreal** and **modred** to reduce the order of a fifth-order randomly generated state-space system.

reduce.m

```
% reduce
% Example of model reduction.
[arand,brand,crand,drand]=rmodel(5);
[abal,bbal,cbal,g,Tbal]=balreal(arand,brand,crand);
fast_states=find(g<g(1)/10);
[amod,bmod,cmod,dmod]=modred(abal,bbal,cbal,drand, ...
fast_states);
if fast_states == [],
     amod=abal; bmod=bbal; cmod=cbal;
     dmod=drand;
end
g
states_removed=length(abal) - length(amod)
original_eigenvalues=eig(abal)
final_eigenvalues=eig(amod)
```

The command **[abal,bbal,cbal,g,Tbal]=balreal(arand,brand,crand)** performs a similarity transformation, **Tbal**, on the given randomly generated fifth-order system (**arand,brand,crand**) such that the transformed system is a balanced realization, that is, a realization for which the controllability and observability grammians are equal and diagonal. The output variable **g** is a vector containing the diagonal elements of the controllability (observability) grammians. The command as shown in code line 4 of *reduce.m* produces a reduced-order state-space model. The input vector **fast_states** has elements equal to the index numbers of the states to be eliminated. In *reduce.m*, we chose to eliminate those states corresponding to very small elements on the diagonal of the grammian. The **find** command is used to identify the index numbers of those components of **g** which are small and to generate the vector **fast_state**. Running *reduce.m* will yield something different each time because the given model is randomly generated. Our output was as follows:

```
g =
    20.8006    4.4021    0.4987    0.1172    0.0030
states_removed =
    3
original_eigenvalues =
        -1.1643 + 0.8411i
        -1.1643 - 0.8411i
        -0.0599 + 0.0620i
        -0.0599 - 0.0620i
        -0.2345
final_eigenvalues =
        -0.0599 + 0.0621i
        -0.0599 - 0.0621i
```

Typically, the final eigenvalues will be different from the original eigenvalues. However, in the preceding example the final eigenvalues are very close to two of the original eigenvalues. This is because the eigenvalues corresponding to the fast states removed were significantly greater than the eigenvalues corresponding to the slow states.

7.3 Design Example

Consider the system of *ssrloc.m* from Chapter 5, which has two poles on the real axis and a complex conjugate pair of lightly damped poles typical of flexible structures. According to the root locus plot for this system, illustrated in Figure 5.3, using a constant gain feedback to control this system can easily lead to instability. In this section we use state-space design techniques to produce more acceptable closed-loop dynamics.

Suppose we would like the closed-loop system to have a pair of dominant poles with a damping coefficient no less than 0.3 and a natural frequency as high as possible. Also assume that the states of our system, described by **a**, **b**, **c**, **d** in file *ssrloc.m*, are not available to the controller. Then, to move the closed-loop poles as desired, we should design a controller and an observer.

As a first try, we design a controller and an observer to move the poles of our system so that the pair of dominant closed-loop poles has a damping coefficient of 0.707 and a natural frequency of 3 rad/s. The file *designexample.m* contains the commands to calculate the controller gain matrix

K, the observer gain matrix **L**, and the state-space model **areg**, **breg**, **creg**, and **dreg** of the final closed-loop system (of Figure 7.3) given a damping coefficient **dmp** and a natural frequency **wn**. Note that we use the **input** command in the script file so that the user can specify the damping and natural frequency at the time the file is run. This is useful because we expect to test several different values on our design procedure.

designexample.m

```
% designexample
% A state-space design to specified damping and natural
% frequency.
dmp=input('What is desired damping ratio?');
wn=input ('What is desired natural frequency?');
a=[-1.5 -13.5 -13 0; 1 0 0 0; 0 1 0 0; 0 0 1 0];
b=[1; 0; 0; 0];
c=[0 0 0 1];
d=0;
[num2,den2]=ord2(wn,dmp);
dominant=roots(den2);
newpoles=[dominant' 10*real(dominant(1))
10*real(dominant(1))];
K=acker(a,b,newpoles)
observerpoles=10*newpoles;
L=acker(a',c',observerpoles)
areg=[(a-b*K) b*K; zeros(size(a)) (a-L'*c)];
breg=[b; zeros(size(b))];
creg=[c zeros(size(c))];
drg=0;
```

The rest of the MATLAB commands in *designexample.m* are very similar to the commands used in the regulator design of Section 7.1 in the file *regulator.m*. However, our current system (*ssrloc.m*) is fourth-order, so we must select four new closed-loop pole locations. We choose the two poles that are not dominant to be a factor of 10 faster than the real part of the selected dominant poles. Additionally, we choose the observer poles to be 10 times faster than the controller poles.

For our first try, we enter the following:

```
>> designexample
What is desired damping ratio?  0.707
What is desired natural frequency?  3.0
```

K =

    1.0e+03 *
    0.0452   0.6253   2.2771   4.0488

L =

    1.0e+07 *
    3.6345   0.2189   0.0063   0.0000

This tells us that we can move the poles as desired. However, according to the preceding MATLAB output we must use very high gains K and L to achieve these desired dynamics. The high magnitudes of the elements of the matrix L are not too serious a problem since the output of the observer feedback is used internally to the controller. On the other hand, the actuator signal $u(t)$ is equal to $K\hat{x}(t)$. Physical actuators usually have a maximum allowable input signal or, equivalently, a limited range. Therefore, an excessively high gain matrix K can lead to actuator saturation. As a reasonable assumption, we suppose that to avoid saturating our actuator we need the elements of K to be less than or equal to 10. This implies that due to physical constraints we cannot achieve 0.707 damping and 3.0 rad/s natural frequency using an observer and controller.

For our second try, we allow the desired damping ratio to drop to 0.3 (the minimum requirement) and retain the desired natural frequency of 3.0 rad/s as follows:

```
>> designexample
What is desired damping ratio?  0.3
What is desired natural frequency?  3.0
K =
      18.3000     108.9000     294.800     729.0000
L =
      1.0e+06 *
      6.6955     0.2872     0.0119     0.0002
```

Again, the gain matrix K is too high for the physical capability of our system. Therefore, we try next for a damping ratio of 0.3 and a natural frequency of 1.0 rad/s.

```
>> designexample
What is desired damping ratio?  0.3
What is desired natural frequency?  1.0
K =
       5.1000     0.1000     -1.6000     9.0000
```

L =

    1.0e+04 *
    5.9327   0.8642   0.1250   0.0065

This time our gain matrix **K** is acceptable, so our design meets the performance requirements. Further design iterations might lead to an acceptable design with a slightly higher natural frequency; however, we will settle for the design just obtained.

A check of the eigenvalues of the corresponding closed-loop system confirms that we have achieved the desired system dynamics.

```
>> damp(areg)
   Eigenvalue              Damping        Freq. (rad/sec)
   -0.3000 + 0.9539i       0.3000         1.0000
   -0.3000 - 0.9539i       0.3000         1.0000
   -3.0000                 1.0000         3.0000
   -3.0000 + 9.5394i       0.3000         10.0000
   -3.0000 - 9.5394i       0.3000         10.0000
   -3.0000                 1.0000          3.0000
   -30.0000 + 0.0000i      1.0000         30.0000
   -30.0000 - 0.0000i      1.0000         30.0000
```

To illustrate the improvements of this state-space design over the design using constant gain feedback, we superimpose in Figure 7.4 the new closed-loop poles onto the original system root locus of Figure 5.3. The four desired closed-loop poles of the state-space design show up on the plot of Figure 7.4 as the poles *not* on the root locus. Two of these poles are located at $s = -3$, although the multiplicity is difficult to see in the figure. The other two (dominant) poles lie on the line of 0.3 damping. Using **sgrid**, we place the line of 0.3 damping ratio on the plot so you can see that the new closed-loop poles are more highly damped than the original lightly damped poles. Of course, since the four closed-loop poles do not lie on the root locus, we could not have generated our final state-space design with constant feedback only. The four observer poles do not show up in the figure because they fall outside the range of the plot. Figure 7.4 was generated by running file *superpose.m* after running *designexample.m* with **dmp** = 0.3 and **wn** = 1.0.

superpose.m

```
% superpose
% Graphical comparison of state-space and root locus
```

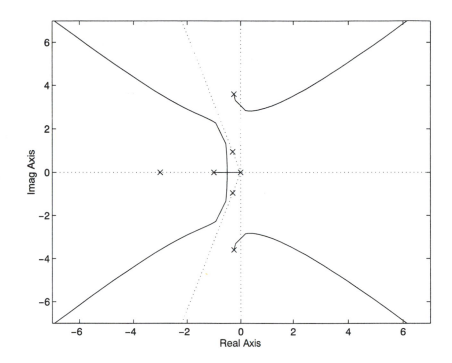

Figure 7.4   Superposition of new closed-loop poles on original root locus.

```
% designs.
clg
rlocus(a,b,c,d);
hold on
pzmap(areg,breg,creg,drg); sgrid(0.3,[])
hold off
```

7.4 Exercises

1. Consider the controller and observer designed in *regulator.m* for $G(s) =$ $1/s(s + 1)(0.2s + 1)$.

   (a) Suppose we want the closed-loop system to have two dominant poles corresponding to a second-order system with a damping ratio of 0.707 and a natural frequency of 3. Now suppose that to ensure that these two poles are dominant, we require the third closed-loop pole to be 20 times the real part of the dominant closed-loop poles. Compute the controller gain **K**. How does this compare to the **K** calculated in *regulator.m*? Explain. How do the dynamics of the new closed-loop system compare to the dynamics of the closed-loop system of *regulator.m*? Explain.

(b) Repeat part (a) assuming that we allow the third closed-loop pole to be only four times the real part of the dominant closed-loop poles.

(c) Let the third closed-loop pole be ten times the real part of the dominant closed-loop poles, as in *regulator.m*. However, now suppose we require that the observer poles be 20 times as great as the desired closed-loop poles. Compute the observer gain **L**. How does this compare to the **L** calculated in *regulator.m*? Explain. How do the dynamics of the new closed-loop system compare to the dynamics of the closed-loop system of *regulator.m*? Explain.

(d) Repeat part (c) assuming that we allow the observer poles to be four times as great as the desired closed-loop poles.

(e) Begin with the original design of *regulator.m*. What is the maximum natural frequency we can achieve if we require that each element of **K** be less than or equal to 10?

(f) Repeat part (e), except suppose that our only requirement is that the damping ratio is at least 0.5.

2. (a) Check the controllability and observability of the following systems.

$$(i)\dot{x} = \begin{bmatrix} 1 & 4 & 3 \\ 0 & 2 & 16 \\ 0 & -25 & -20 \end{bmatrix} x + \begin{bmatrix} -1 \\ 0 \\ 0 \end{bmatrix} u, \quad y = [-1 \ 3 \ 0]x$$

$$(ii)\dot{x} = \begin{bmatrix} 1 & 0 & 0 \\ 0 & 0 & 0 \\ -2 & -4 & -3 \end{bmatrix} x + \begin{bmatrix} 1 \\ 0 \\ -1 \end{bmatrix} u, \quad y = [1 \ 0 \ 0]x$$

(b) For each of the systems of part (a), find the controllable modes, uncontrollable modes, observable modes, and unobservable modes. (Hint: Use **ctrbf** and **obsvf**.)

(c) For each of the systems of part (a), assuming the states are available for feedback, determine whether or not a state-feedback controller can be designed to stabilize the system.

(d) For each of the systems of part (a), assuming the states are *not* available for feedback, determine whether or not a controller together with an observer can be designed to stabilize the system.

3. Assuming the states are available for feedback, design a controller that will stabilize the system (**A, B, C, D**) of *modalform.m*.

4. For the plot of the state-space design in Figure 7.3, we eliminated steady-state error by scaling the input to the system by a fixed gain. This is not the most reliable method for eliminating steady-state error because it assumes perfect knowledge of the system. Adding integral control is a much more effective method. Assume the states of the system for this example (*regulator.m*) are available. Design a controller with integral control that achieves the desired performance (as in *regulator.m*) as well as zero steady-state error. [Hint: Feed back the output $y(t)$ in Figure 7.1 and integrate the error e_I where $e_I(t) = r(t) - y(t)$. Let x_N be the new state corresponding to the output of the integrator, i.e., $\dot{x}_N(t) = e_I(t)$. Incorporate this new state into the state-space model of the closed-loop system by augmenting the state to be

$$\begin{bmatrix} x \\ x_N \end{bmatrix}$$

Let

$$u(t) = -Kx - K_I x_N = -\begin{bmatrix} K & K_I \end{bmatrix} \begin{bmatrix} x \\ x_N \end{bmatrix}$$

where K is the controller gain as before and K_I is the integrator gain.]

8 DISCRETE-TIME CONTROL SYSTEMS

The study of digital control systems deals with the effect on system dynamics of finite word length as well as discretization of time in digital computation. However, because digital control is such a large topic and because MATLAB is best suited for investigating system effects due to the discretization of time, we focus in this chapter on discrete-time systems rather than on the effects of finite word length. In particular, we emphasize discrete-time systems that are obtained by exact sampling of continuous-time systems, often referred to as "sampled-data" systems.

Furthermore, since the subject of discrete-time control systems itself is extensive, we have limited ourselves to three topics that are related to the previous material in this book. These are the conversion of continuous-time systems to equivalent discrete-time systems, the effects of sampling on system behavior, and the discrete-time versions of root locus, Nyquist, Nichols, and Bode plots. We recommend that the reader searching for a deeper understanding of digital (or discrete-time) control systems consult *Digital Control of Dynamic Systems*, second edition, by Franklin, Powell, and Workman (Addison-Wesley Publishing Co., 1990). This reference is particularly useful in learning to use MATLAB because an earlier version of the book was the basis for the MATLAB routines currently available for studying discrete-time systems.

8.1 Creating Discrete-Time Systems

Discrete-time systems can be entered into MATLAB in two distinct ways.

If you know the discrete-time system description, you can input it directly into MATLAB in either state-space or transfer function form. Because many discrete-time systems arise from the sampling or discretization of continuous-time systems you may instead start with a continuous-time system. MATLAB makes it possible for you to enter the continuous-time system (as done in previous chapters) and use MATLAB to discretize the time. The important MATLAB commands for this purpose are **c2d**, **c2dt**, and **c2dm**.

Our emphasis will be on discretization of continuous-time systems. However, we will begin by describing methods for inputting discrete-time systems directly into MATLAB. A discrete-time system can be described in the state space by

$$x^*(k+1) = Ax^*(k) + bu^*(k)$$
$$y^*(k) = cx^*(k) + du^*(k)$$

where the superscript * indicates a discrete-time signal and

$x^*(k)$ is the state vector at "index" k

$y^*(k)$ is the output at "index" k

$u^*(k)$ is the input at "index" k

the index k takes values $0, 1, 2, \ldots$

or by its transfer function (the z-transform of its discrete-time impulse response) as

$$G(z) = \frac{K(z^m + a_1 z^{m-1} + \cdots + a_m)}{(z^n + b_1 z^{n-1} + \ldots + b_n)}$$

Entering either of these descriptions in MATLAB is virtually the same as entering the corresponding description for a continuous-time system. The only difference is that if you want to display a discrete-time transfer function, you need to include **'z'** as the last input argument in the **printsys** command. This is illustrated in *discretesys.m.*

discretesys.m
```
% discretesys
% This creates and displays discrete-time systems.
A=[0 1 0 0; 0 0 1 0; 0 0 0 1; -1 -1 -1 -1];
b=[0;0;0;1];
c=[1 0 0 0];
d=0;
```

```
printsys(A,b,c,d)
num=[1 1 1];
den=[1 3 5 7];
printsys(num,den,'z')
```

The results of running the file are

```
a =
          x1          x2          x3          x4
    x1    0           1.00000     0           0
    x2    0           0           1.00000     0
    x3    0           0           0           1.00000
    x4    -1.00000    -1.00000    -1.00000    -1.00000
b =
          u1
    x1    0
    x2    0
    x3    0
    x4    1.00000
c =
          x1          x2          x3          x4
    y1    1.00000     0           0           0
d =
          u1
    y1    0
num/den =

                    z^2 + z + 1

          - - - - - - - - - - - -

          z^3 + 3 z^2 + 5 z + 7
```

We will describe ways to analyze these systems shortly. First, we describe the MATLAB commands for discretizing a continuous-time system. The situation of primary interest is illustrated in Figure 8.1. We are given a continuous-time plant, $G_p(s)$, which we intend to control by means of a discrete-time controller, $G_c(z)$. We assume, as is true in practice, that the discrete-time controller produces an output signal every T_s seconds and we let $u^*(k)$ be equivalent to $u(t = kT_s)$ as shown in Figure 8.2. Such a discrete signal will not drive a continuous-time system. Thus, it is necessary to convert $u^*(k)$ to a continuous-time signal. This is commonly done by means of either a zero-order hold (zoh) or a first-order hold (foh). Similarly, the discrete-time controller cannot accept continuous-

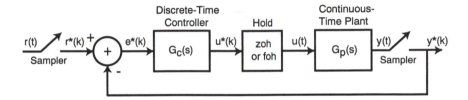

Figure 8.1 Implementation of a discrete-time controller for a continuous-time plant.

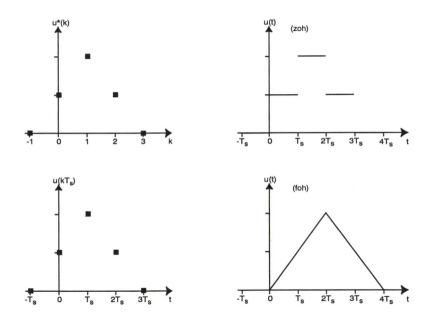

Figure 8.2 Relation between a discrete signal $u^*(k)$ (upper left), the corresponding discrete-time signal $u(kT_s)$ (lower left), the output $u(t)$ that results when $u^*(k)$ is input to a zero-order hold (upper right), and the output $u(t)$ that results when $u^*(k)$ is input to a first-order hold (lower right).

time inputs so we sample the output, $y(t)$, of the continuous-time plant every T_s seconds. The resulting discrete-time signal $y^*(k)$ is equivalent to $y(t = kT_s)$. Thus the system that takes input $u^*(k)$ into output $y^*(k)$ is a discrete-time system that is equivalent to the continuous-time system denoted by $G_p(s)$.

The simplest discretization case to analyze is for a continuous-time system given in state-space form with continuous-time input $u(t)$ generated from a discrete-time input $u^*(k)$ followed by a zero-order hold. In this case, we can derive an exact formula for $x^*(k + 1)$, the state at time $(k + 1)T_s$, in terms of $x^*(k)$ and $u^*(k)$, the state and control at time kT_s, where T_s is the sampling period.

Consider the continuous-time system

$$\dot{x}(t) = Ax(t) + bu(t), \qquad x(t_0) = x_0 \tag{8.1}$$

$$y(t) = cx(t) + du(t)$$

We assume $t_0 = kT_s$, so $x_0 = x(kT_s)$, and $u(t) = u(kT_s)$ for $kT_s \le t < (k + 1)T_s$.

Integrating Eq. (8.1) gives

$$x((k + 1)T_s) = e^{AT_s}x(kT_s) + \int_{kT_s}^{(k+1)T_s} e^{A(k+1)T_s}e^{-A\tau}b \, d\tau \, u(kT_s)$$

$$= e^{AT_s}x(kT_s) + \int_0^{T_s} e^{AT_s}e^{-A\sigma}d\sigma \, bu(kT_s)$$

Thus,

$$x((k + 1)T_s) = A_d x(kT_s) + b_d u(kT_s) \tag{8.2}$$

$$A_d = e^{AT_s}$$

$$b_d = \int_0^{T_s} e^{AT_s}e^{-A\sigma}d\sigma b$$

It is then customary for notational purposes to drop the T_s in our specification of the discrete-time system and rewrite Eq. (8.2) as

$$x^*(k + 1) = A_d x^*(k) + b_d u^*(k) \tag{8.3}$$

$$y^*(k) = cx^*(k) + du^*(k)$$

MATLAB implements this formula by means of the command **c2d** [continuous **(c)** to **(2)** discrete **(d)**]. We illustrate the use of this command with file *simplec2d.m*. In this file we enter a continuous-time state-space model (**A, b, c, d**) and use **c2d** to compute A_d and b_d from Eq. (8.3) for a sampling interval **Ts**=0.1 and store them in variables **Ad** and **bd**. Notice that the c and d matrices are not needed as input to **c2d**. This should not be surprising, since c and d are unchanged by the discretization.

simplec2d.m
```
% simplec2d
% A simple continuous to discrete conversion.
```

```
A=[0 1 0 0; 0 0 1 0; 0 0 0 1; -1 -1 -1 -1];
b=[0;0;0;1];
c=[1 0 0 0];
d=0;
Ts=0.1;
[Ad,bd]=c2d(A,b,Ts)
```

The result of running the file is

```
Ad =
    1.0000      0.1000      0.0050      0.0002
   -0.0002      0.9998      0.0998      0.0048
   -0.0048     -0.0050      0.9950      0.0950
   -0.0950     -0.0998     -0.1000      0.9000
bd =
    0.0000
    0.0002
    0.0048
    0.0950
```

MATLAB also contains a slightly more complicated command **c2dt** for discretizing continuous-time systems that have a delay in the input, that is, continuous systems of the form:

$$\dot{x}(t) = Ax(t) + bu(t - \lambda)$$
$$y(t) = cx(t)$$

where λ is the time delay. In file *notsosimplec2d.m* we enter the same system description as in the previous example except we include a time delay $\lambda = 0.51$, which we store in the variable **lambda**. Using the command **c2dt**, we can get MATLAB to give us A_d, b_d, c_d, d_d for the corresponding discretized system:

$$x^*(k+1) = A_d x^*(k) + b_d u^*(k)$$
$$y^*(k) = c_d x^*(k) + d_d u^*(k)$$

notsosimplec2d.m
```
% notsosimplec2d
% A continuous to discrete conversion that includes a
% delay.
A=[0 1 0 0; 0 0 1 0; 0 0 0 1; -1 -1 -1 -1];
b=[0;0;0;1];
```

```
c=[1 0 0 0];
d=0;
Ts=0.1;
lambda=0.51;
[Ad,bd,cd,dd]=c2dt(A,b,c,Ts,lambda)
```

The result of running file *notsosimplec2d.m* is

```
Ad =
  Columns 1 through 5
    1.0000      0.1000      0.0050      0.0002      0.0000
   -0.0002      0.9998      0.0998      0.0048      0.0000
   -0.0048     -0.0050      0.9950      0.0950      0.0009
   -0.0950     -0.0998     -0.1000      0.9000      0.0091
    0           0           0           0           0
    0           0           0           0           0
    0           0           0           0           0
    0           0           0           0           0
    0           0           0           0           0
    0           0           0           0           0
  Columns 6 through 10
  0.0000      0.0000      0           0           0
  0.0001      0.0000      0           0           0
  0.0039      0.0000      0           0           0
  0.0860      0.0000      0           0           0
  1.0000      0           0           0           0
  0           1.0000      0           0           0
  0           0           1.0000      0           0
  0           0           0           1.0000      0
  0           0           0           0           1.0000
  0           0           0           0           0
```

bd =

$$
\begin{array}{c}
0 \\
0 \\
0 \\
0 \\
0 \\
0 \\
0 \\
0 \\
0 \\
1
\end{array}
$$

cd =

          1 0 0 0 0 0 0 0 0 0

dd =

          0

Notice that MATLAB has added states to our fourth-order system to account for the delay. In general, a delay λ will increase the state dimension by the smallest integer $m \geq \lambda/T_s$, where T_s is the sampling interval (see Franklin, Powell, and Workman). The delay in our example is 5.1 times the sampling interval, T_s, so MATLAB adds six states to the original four states for a total of ten states.

In addition to the zoh technique, MATLAB makes available other methods for converting continuous-time systems to discrete-time systems. Using **c2dm** (where the **m** stands for "method"), you can select which discretization method MATLAB should use. For state-space models where you have specified a system **a, b, c, d** and a sampling interval **Ts**, use

[ad, bd, cd, dd] = c2dm(a,b,d,c, 'method')

to compute the equivalent discrete-time system **ad, bd, cd**, and **dd** with the method indicated in the last input argument.

Similarly, you can use **c2dm** to convert a continuous-time system specified by a transfer function $G(s)$ to its discrete-time equivalent $G_d(z)$. In this case, where you have specified the numerator and denominator polynomials of $G(s)$ as **num** and **den** and the sampling interval as **Ts**, use the command **[numd,dend]=c2dm(num,den,Ts,'method')** to compute the numerator and denominator polynomials, **numd** and **dend**, of $G_d(z)$ with the method specified.

The discretization methods available with **c2dm** include (1) the zero-order-hold method implemented with **'zoh'** as the last input argument to **c2dm**; (2) the first-order-hold method implemented with **'foh'**; (3) Tustin's method (also known as the bilinear transformation) implemented with **'tustin'**; (4) Tustin's method with frequency prewarping implemented with **'prewarp'**; and (5) the matched pole-zero method implemented with **'matched'**. An example of each of these methods is provided in the following.

In our first example in the file *c2dmexample1.m*, we demonstrate using **c2dm** to perform a discretization by the zoh method of a system specified by a transfer function. We first create a transfer function for a second-order system with a lightly damped pair of complex conjugate poles at $-0.1 \pm i$. This system has an approximate bandwidth of 1 rad/s. To decide on a sampling interval for the discretization, we use a theorem due to Nyquist. This theorem says that it is possible to reconstruct exactly a continuous-time signal $y(t)$ from its sampled equivalent $y(kT_s)$ provided the sampling frequency $2\pi/T_s$ is twice the bandwidth of the Fourier transform of $y(t)$. If we denote this bandwidth by ω_B, then the Nyquist rate, $2\omega_B$, is the minimum sampling frequency that allows perfect reconstruction. Since the transfer function $G(j\omega)$ of a system is just the Fourier transform of its impulse response, we can apply this result directly to transfer functions.

The Nyquist rate of the example system is then 2 rad/s. Thus, a sampling interval of $T_s = 2\pi/2 \approx 3$ seconds would be adequate if the conditions of the theory were exactly satisfied. However, the theory assumes that the transfer function is zero for frequencies outside the bandwidth. This is often not true and is certainly not true in our example. As a result, it is common to choose a sampling frequency ten times the Nyquist rate in order to ensure that the continuous-time signal can be accurately reconstructed. Thus, we choose $T_s = 2\pi/(10)(2\omega_B) = 0.3$ second.

After specifying the sampling interval **Ts**, we then use the command **c2dm** as described to discretize the system. Because the zoh method is the default for **c2dm** we need not specify the method as an input argument. Using the **printsys** command, we then display $G_d(z)$ for comparison with $G(s)$. Finally, we use the command **step** to show the **step** response of the continuous-time system, and we introduce the command **dstep** to display the step response of the discrete-time system. The command **dstep** for discrete-time systems is equivalent to **step** for continuous-time systems. The results are shown immediately after the file and in Figure 8.3. Be sure to notice that the horizontal axis in the graph of the discrete-time

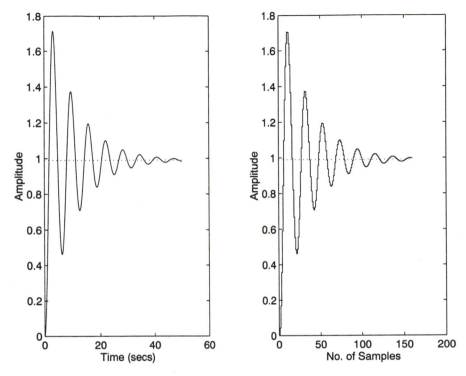

Figure 8.3 Left: Step response of the continuous-time system having transfer function $G(s) = 1/(s + 0.1 + j)(s + 0.1 - j)$. Right: Step response of the corresponding discrete-time system with sampling interval $T_s = 0.3$ second (conversion by zoh).

response is the number of samples rather than time.

c2dmexample1.m

```
% c2dmexample1
% An example of continuous to discrete conversion using
% zoh.
[num,den]=zp2tf([],[-0.1+j    -0.1-j],1);
Ts=.3;
[numd,dend]=c2dm(num,den,Ts);
clf
printsys(num,den,'s')
printsys(numd,dend,'z')
subplot(121),   step(num,den)
subplot(122),   dstep(numd,dend)
```

num/den =

$$\frac{1.000}{s\hat{\ }2 + 0.2\ s + 1.01}$$

num/den =

$$\frac{0.04378\ z + 0.04291}{z\hat{\ }2 - 1.854\ z + 0.9418}$$

Our second example, *c2dmexample2.m*, uses the same $G(s)$ as the previous example and the same sampling interval, $T_s = 0.3$ second. We compute the corresponding $G_d(z)$ by means of the four other methods available. The method is specified each time by the fourth argument of **c2dm**. When the method is **'prewarp'** a fifth argument is needed. The fifth argument must be a frequency $\omega_1 < \pi/T_s$ as explained in Franklin, Powell and Workman (p. 140). We chose $\omega_1 = 1$ simply as a demonstration. The results for each method are shown immediately after the file. We discuss the different methods after the results.

c2dmexample2.m

```
% c2dmexample2
% Continuous to discrete conversion using 4 different
% methods.
[num,den]=zp2tf([],[-0.1+j     -0.1-j],1);
Ts=.3;
[numd,dend]=c2dm(num,den,Ts,'tustin');
printsys(numd,dend,'z')
[numd,dend]=c2dm(num,den,Ts,'foh');
printsys(numd,dend,'z')
[numd,dend]=c2dm(num,den,Ts,'prewarp',1);
printsys(numd,dend,'z')
[numd,dend]=c2dm(num,den,Ts,'matched');
printsys(numd,dend,'z')
```

num/den =

$$\frac{0.02137\ z\hat{\ }2 + 0.4275\ z + 0.02137}{z\hat{\ }2 - 1.857\ z + 0.943}$$

num/den =

$$\frac{0.0141z^2 + 0.05871z + 0.01428}{z^2 - 1.854z + 0.9418}$$

num/den =

$$\frac{0.02169\ z^2 + 0.04337\ z + 0.02169}{z^2 - 1.854\ z + 0.9426}$$

num/den =

$$\frac{0.04335z + 0.4335}{z^2 - 1.854\ z + 0.9418}$$

Notice that each method gives a different answer for $G_d(z)$ when applied to our simple example. To understand why this is so, it is important to understand first that, in general, each discretization method is based on a different interpretation of the equivalence between the continuous- and discrete-time system descriptions. For example, for discretization by the zoh method we showed that the discrete- and continuous-time systems have identical outputs at each sampling instant. That is, letting $y(t = kT_s)$ denote the output of the continuous-time system at $t = kT_s$ and letting $y^*(k)$ denote the output of the discrete-time system at sample instant k (remember, sample instant k corresponds to $t = kT_s$) implies $y(t = kT_s) = y^*(k)$. Additionally, the zoh method assumes that the inputs to the two systems are the same: $u(\ell T_s \leq t < (\ell + 1)T_s) = u^*(\ell)$ for all $\ell \leq (k - 1)$.

A second example is the first-order hold, or foh method, which is an implementation of the "triangle hold" method described in Franklin, Powell, and Workman. This method is similar to the zoh method except that instead of holding the input signal constant between sampling instants, the foh method causes it to vary linearly between values of $u^*(k)$. That is, the discrete input signals $u^*(k)$ are connected by straight lines. This provides a continuous-time input signal $u(t)$ to the continuous-time plant.

As you might guess, because straight-line interpolation more accurately approximates the input signal than a constantly held input signal, the foh method often produces a more accurate discrete-time approximation to the continuous-time system than the zoh method.

The sense in which the zoh and foh methods produce discrete-time systems that are equivalent to continuous-time systems is easy to appreciate. However, the conversion of a $G(s)$ into a $G_d(z)$ by means of the zoh or foh does not lend itself to hand calculation in any but very simple cases. Tustin's method, on the other hand, directly converts $G(s)$ into $G_d(z)$ by the following simple substitution.

Let

$$s = \frac{2}{T_s}\left(\frac{z-1}{z+1}\right)$$

Then

$$G_d(z) = G\left(\frac{2}{T_s}\frac{(z-1)}{(z+1)}\right)$$

Thus, Tustin's method (also known as the bilinear transformation method) is relatively convenient if you don't have a computer handy. Tustin's method is also very convenient for converting analog controller designs into digital ones. For example, suppose we have designed a control system for a plant $G_p(s)$. Suppose further that our controller is $G_c(s)$. We can create a digital equivalent to $G_c(s)$, which we will call $G_{cd}(z)$, by Tustin's method. Typically, we would do this in order to implement the controller digitally on a computer. If the sampling interval is small enough, the input-output behavior of the closed-loop system with the digital controller $G_{cd}(z)$ should be virtually indistinguishable from the behavior of the closed-loop system with the continuous controller $G_c(s)$. We will not discuss **'prewarp'** and **'matched'**. See Franklin, Powell, and Workman for further information on these methods.

We conclude this section by mentioning that there are three MATLAB commands to convert discrete-time systems to equivalent continuous-time systems. They are **d2c**, **d2ct**, and **d2cm**. They work as algorithmic inverses of the corresponding continuous-to-discrete algorithms and use the same syntax. We leave them for you to explore on your own.

8.2 Time-Domain and Pole-Zero Analysis of Discrete-Time systems

All of the usual tools for analyzing and designing continuous-time systems have discrete-time equivalents. In this section we consider time-domain analysis techniques such as pulse responses (the discrete-time version of impulse responses) and pole-zero analysis techniques for discrete-time systems. Frequency-domain methods are considered in the next section.

It should be no surprise that discrete-time systems have impulse, step, and initial condition responses. In MATLAB these responses can be produced with **dimpulse**, **dstep**, and **dinitial**, respectively. Similarly, **dlsim** (analogous to **lsim**) can be used to compute the discrete-time system response to an arbitrary input. Comparing the continuous- and discrete-time system step responses in Figure 8.3, it should be obvious that specifications like time to peak, peak overshoot, and settling time apply equally well to both systems. We can also define natural frequencies and damping rat for second-order discrete-time systems. Additionally, the numerator d denominator polynomials of a given system description $G(z)$ can be f ctored to obtain the system poles and zeros in the z-plane. Thus, all the elements of the root locus plot are also available for discrete-time syste ns. The associated commands **ddamp** and **zgrid** will be introduced in th section.

We h ve already shown how to use **dstep** to create step responses for discrete-time systems. We can just as easily compute time-domain performance measures for discrete-time systems as we did for continuous-time systems in Chapter 4 with the file *stepanalysis.m*. Because MATLAB solves for the step response of continuous-time systems by discretizing time, the file *stepanalysis.m* of Chapter 4 will correctly compute time to peak, percent overshoot, rise time, and settling time for discrete-time systems provided three minor changes are made. The first line of the file must be changed so that the specified poles describe a stable discrete-time system of interest. The values given in the file, $-1 \pm 3i$, are outside the unit circle, so a discrete-time system with these poles would be unstable. The second line, which computes the final value, must also be changed. The correct formula for the final value of a signal $x^*(k)$ having z-transform $X(z)$ is (see Nise, Chapter 13)

$$\lim_{k \to \infty} x(k) = \lim_{z \to 1}(1 - z^{-1})X(z)$$

This can be implemented by replacing the second line of *stepanalysis.m* by

```
finalvalue = polyval(conv([1 -1],num),1)/polyval ...
(conv([1 0],den),1)
```

Finally, **step** must be changed to **dstep** in the third line.

In the next example, file *dimpulseex.m*, we demonstrate **dimpulse**, which computes the pulse response for a discrete-time system. The command

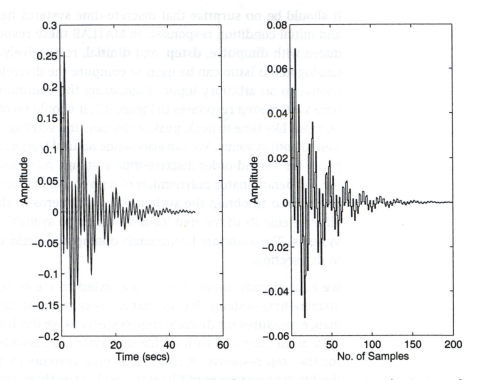

Figure 8.4   Left: Impulse response of the continuous-time system having transfer function $G(s)=(s+0.1+2i)(s+0.1-2i)/(s+0.1+i)(s+0.1-i)(s+0.1-5i)$ $(s+0.1+5i)$. Right: Pulse response of the corresponding discrete-time system with sampling interval $T_s=0.3$ second (conversion by zoh).

dimpulse for discrete-time systems is equivalent to **impulse** for continuous-time systems. We begin by creating a continuous-time system with lightly damped complex conjugate poles at $-0.1 \pm j$ and $-0.1 \pm 5j$. The system also has complex conjugate zeros at $-0.1 \pm 2j$. We plot the impulse response of this system using **impulse** in the left-hand plot of Figure 8.4. You can see that the poles at $-0.1 \pm 5j$ superpose a very lightly damped oscillation on the response due to the lower-frequency poles.

dimpulseex.m

```
% dimpulseex
% Computing the impulse response of a discrete-time
% system obtained by discretizing a continuous-time
% system.
zros=[-.1+2*i  -.1-2*i];  %zeros is a MATLAB function
poles=[-.1+i  -.1-i   -.1+5*i  -.1-5*i];
K=1;
Ts=0.3;
```

```
[num,den]=zp2tf(zros',poles',K);
[numd,dend]=c2dm(num,den,Ts,'zoh');
clf
subplot(121)
impulse(num,den)
subplot(122)
dimpulse(numd,dend)
```

Next we consider the discrete-time version of this system. To do so we must first select a sampling interval T_s. If we take $\omega_B = 5$ rad/s as the bandwidth of this continuous-time system, then the Nyquist rate is 10 rad/s and any sampling interval less than $2\pi/10$ rad/s = 0.6 second should provide enough samples to permit reconstruction of the responses of the original continuous-time system. A better choice for the bandwidth would have been $\omega_B = 7$ rad/s as can be seen in Figure 8.12. Remember that the reconstruction will not be perfect because our example system will actually produce a nonzero response to inputs at frequencies higher than 5 rad/s. We create a discrete-time version of our system using **c2dm**, specifying $T_s = 0.3$ second and that the zoh method be used. We then produce the right-hand plot of Figure 8.4 by means of **dimpulse**. As you can see from Figure 8.4, the pulse response of the discrete-time system looks very similar to the impulse response of the continuous-time system. Note, however, that the amplitudes of the two responses are not the same. It is generally necessary to scale the amplitudes of the impulse responses of discrete-time systems by $1/T_s$ in order to make them match the amplitudes of the impulse responses of the corresponding continuous-time systems.

In our next example, *samplingex.m*, we investigate discrete-time system pulse responses as a function of sampling interval. First, we create a continuous-time system that is predominantly second-order with dominant poles at $-0.25 \pm 0.9682i$ and a bandwidth of approximately $\omega_B = 1$ rad/s. The system is actually third-order. The third pole is at -10, much faster than the dominant poles. The system also has zeros at $-0.1 \pm 1.4107i$ and a gain of 10. We then plot the impulse response of the continuous-time system in the upper left-hand corner of Figure 8.5. To facilitate later comparisons, we use a third input argument, **t**, to **impulse** in order to fix the final time of the plot at **Tf** = 13 seconds. We then use a **while** loop to create three different discrete-time approximations to the continuous-time system. Each of these discretizations uses the zoh method; however, each uses a different sampling interval **Ts**. Specifically,

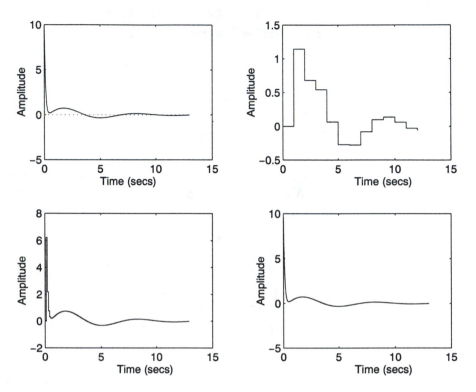

Figure 8.5   Upper left: Impulse response of the continuous-time system having transfer function $G(s) = 10(s^2 + 0.2s + 2)/(s^2 + 0.5s + 1)(s + 10)$. Upper right: Pulse response of the corresponding discrete-time system with sampling interval $T_s = 1$ second (conversion by zoh). Lower left: Same as upper right plot except $T_s = 0.1$ second. Lower right: Same as upper right plot except $T_s = 0.01$ second.

we use sampling intervals of 1 second, 0.1 second, and 0.01 second.

The dominant second-order system has bandwidth $\omega_B = 1$ rad/s, Nyquist rate $2\omega_B = 2$ rad/s, and maximum sampling interval for accurate reconstruction of the continuous-time output signal $T = 2\pi/2 = 3.14$ seconds. Thus, a sampling interval of 1 second should be adequate if we ignore the effects of the pole at –10. The bandwidth of the complete third-order system is $\omega_B = 10$ rad/s. As a result, a sampling interval of 0.1 second should capture nearly all the features of the third-order continuous-time system. However, the continuous-time system does produce nonzero responses to inputs at frequencies above $\omega = 10$ rad/s, so $T_s = 0.01$ second is an even better choice of sampling interval.

samplingex.m
```
% samplingex
% The effect of sampling time on the discrete-time
```

```
% impulse response resulting from discretizing a
% continuous-time system.
num=10*[1 0.2 2];
den=conv([1 0.5 1], [1 10]);
clf
subplot(221)
Tf=13;
t=[0:0.1:Tf];
impulse(num,den,t)
m=1;
while m<=3,
      Ts=1/10^(m-1);
      subplot(221+m)
      [numd,dend]=c2dm(num,den,Ts);
      [y,x]=dimpulse(numd,dend,Tf/Ts);
      t1=[0:Ts:Tf-Ts];
      stairs(t1,y/Ts)
      xlabel('Time(secs)')
      ylabel('Amplitude')
      m=m+1;
end
subplot(111)
```

In the interest of illustrating the comparison between the impulse response of the continuous-time system and the pulse responses of the three discrete-time approximations, we use **dimpulse** in a more sophisticated way than in the previous file. First, we specify the number of sample points **Tf/Ts** at which to compute the response so the final time for each pulse response calculation will always be **Tf**. Second, we specify [**y,x**] as output arguments of **dimpulse**. This causes **dimpulse** to repress the plot and to store the pulse response in the vector **y** and the state vector pulse response in **x** (for the MATLAB selected state-space realization). We then use **stairs** to create a stair step plot of the pulse response. Notice that the first input argument to **stairs** is t1, which is a vector of sampling instants with sampling interval equal to **Ts**. The second argument is the system response $y^*(k)$ scaled by 1/**Ts**. The **stairs** command produces a plot of $y(t)/T_s$ versus time t where $y(t)$ is the zero-order-hold equivalent of $y^*(k)$, that is, $y(t) = y^*(k)$ for $kT_s \leq t < (k+1)T_s$.

The results are the plots shown in Figure 8.5. As you can see, the pulse response of the discretization with sampling interval **Ts** = 1 second (shown in the upper right corner of Figure 8.5) completely misses the large initial jump due to the pole at –10 but follows the slower portion of the response fairly well. The response of the discretization with sampling interval **Ts** = 0.1 second (lower left corner of Figure 8.5) follows the complete response quite well, but it does not quite track the initial peak. The response of the discretization with sampling at **Ts** = 0.01 second duplicates the continuous-time impulse response.

The problem of defining natural frequencies and damping ratios for discrete-time systems, and then computing them, is harder than producing step and pulse responses. A natural approach, and the one followed in MATLAB, is to start with the definition of damping ratio and natural frequency for a continuous-time system. Based on this damping ratio and natural frequency, we can derive equations for the poles of the equivalent discrete-time system. Thus, given poles of a discrete-time system, we can use these equations to compute equivalent damping ratio and natural frequency. Damping ratio and natural frequency are defined for continuous-time systems $H(s)$ of the form

$$H(s) = \frac{\omega_n^2}{s^2 + 2\zeta\omega_n s + \omega_n^2}$$

such that ζ is the damping ratio and ω_n is the natural frequency. These systems have impulse response

$$h(t) = \begin{cases} e^{-at} \sin \omega_c t & t \geq 0 \\ 0 & t < 0 \end{cases}$$

where

$$a = \zeta\omega_n \tag{8.4}$$

$$\omega_c = \omega_n\sqrt{1 - \zeta^2}, \qquad 0 \leq \zeta \leq 1 \tag{8.5}$$

If we sample this impulse response with sampling interval T_s we obtain

$$h^*(k) = \begin{cases} e^{-aT_s k} \sin \omega_c T_s k & k \geq 0 \\ 0 & k < 0 \end{cases}$$

The z-transform of $h^*(k)$ can be computed fairly easily or found in virtually any table of z-transforms (e.g., see Nise, Chapter 13). It is

$$H(z) = \frac{ze^{-aT_s} \sin \omega_c T_s}{z^2 - 2ze^{-aT_s} \cos \omega_c T_s + e^{-2aT_s}}$$

Factoring the denominator polynomial, and writing the factors in polar form, shows that the poles of $H(z)$ are

$$p_1 = re^{j\theta} = e^{-aT_s} e^{j\omega_c T_s}, \qquad p_2 = re^{-j\theta} = e^{-aT_s} e^{-j\omega_c T_s}$$

Using Eqs. (8.4) and (8.5) gives

$$\theta = \omega_n T_s \sqrt{1 - \zeta^2} \tag{8.6}$$

$$r = e^{-\zeta \omega_n T_s} \tag{8.7}$$

Noting that ω_n and T_s always appear as a product and that $\omega_n T_s$ is dimensionless, we can lump $\omega_n T_s$ together. If we hold $\omega_n T_s$ constant, vary ζ from 0 to 1 in steps of 0.1, compute the corresponding values of r and θ, and create a polar plot of r versus θ, then we will have a z-plane contour of constant $\omega_n T_s$. We use a polar plot because r and θ are the polar form for the poles. This is the exact analog of the s-plane circle of constant ω_n as plotted by **sgrid** in the sense that the pulse response of a discrete-time second-order system with poles at $re^{j\theta}$ and $re^{-j\theta}$ where

$$r = e^{-\zeta \omega_n T_s}, \qquad \theta = \omega_n T_s \sqrt{1 - \zeta^2}$$

will exactly match the impulse response of a continuous-time second-order system with damping ratio ζ and natural frequency ω_n at the time instants kT_s.

The first half of file *zgridtest.m* computes a contour of constant $\omega_n T_s$ by implementing the calculation just described.

zgridtest.m

```
% zgridtest
% Demonstration of constant damping ratio & natural
% frequency contours for discretized systems.
% Produce contour of constant omegaTs=pi.
omegaTs=pi;
zeta=[0:.1:1];
theta=omegaTs*sqrt(1-zeta.^2);
r=exp(-zeta*omegaTs);
clf
subplot(121)
polar(theta,r)
% Produce contour of constant zeta=0.7.
zeta=.7;
omegaTs=[0:.1:pi/sqrt(1-zeta^2)];
theta=omegaTs*sqrt(1-zeta^2);
r=exp(-zeta*omegaTs);
subplot(122)
polar(theta,r)
```

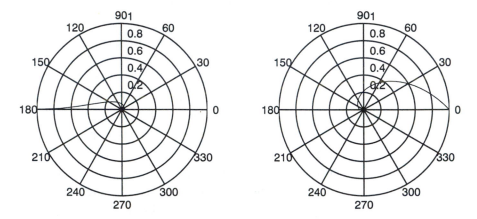

Figure 8.6   Left: Contour of constant $\omega_n T_s = \pi$. Poles of a discrete-time system that lie on this contour correspond to poles of a continuous-time system having natural frequency $\omega_n = \pi/T_s$. Right: Contour of constant $\zeta = 0.7$. Poles of a discrete-time system that lie on this contour correspond to poles of a continuous-time system having damping ratio $\zeta = 0.7$.

We first set $\omega_n T_s = $ **omegaTs** $ = \pi$. We then implement Eqs. (8.6) and (8.7) and create the polar plot using the command **polar**. The result is shown on the left-hand side of Figure 8.6. Similarly, if we hold ζ constant, vary $\omega_n T_s$ from 0 to $\pi/\sqrt{1-\zeta^2}$ in increments of 0.1, and create a polar plot of the resulting r versus θ, then we will have a z-plane contour of constant ζ, that is, the exact analog of an s-plane line of constant ζ as created by **sgrid**. We demonstrate this in the second half of *zgridtest.m*. For this case $\zeta = $ **zeta** $ = 0.7$ and the plot is shown on the right-hand side of Figure 8.6.

The MATLAB command **zgrid** draws the lines of constant damping and natural frequency for you for discrete-time systems analagously to **sgrid**. We illustrate **zgrid** as part of a simple root locus computation in *zrootlocus.m*. The first four lines of the file specify zeros at ±0.3, poles at 0.1 and 0.5±0.5j, and a gain equal to one. The next line creates a transfer function $G(s)$ out of these zeros, poles, and gain.

We then create a root locus plot for the closed-loop system shown in Figure 8.7, where k is the variable gain. We have deliberately drawn a continuous-time feedback loop to emphasize that, to this point in *zrootlocus.m*, we could equally well be describing either a continuous- or discrete-time system. Only **zgrid**, the last command in the file, is specific to discrete-time systems. The command **zgrid** creates a unit circle and constant damping and natural frequency grid lines for discrete-time systems.

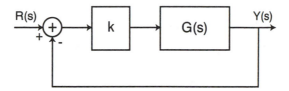

Figure 8.7 Generic proportional unity feedback control system for root locus calculations.

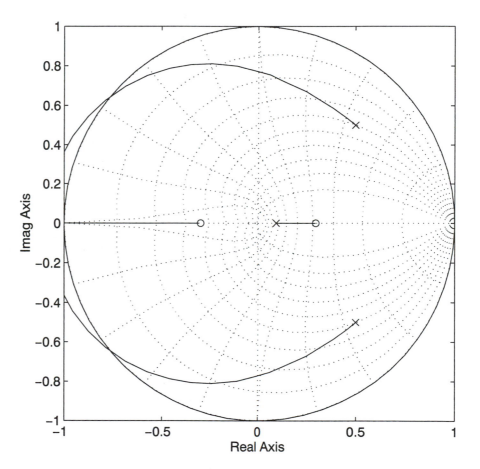

Figure 8.8 Root locus plot for the discrete-time system having transfer function $G(z) = (z + 0.3)(z - 0.3)/(z - 0.1)(z - 0.5 + .5i)(z - 0.5 - 0.5i)$. The unit circle and contours of constant $\omega_n T_s$ and ζ are shown.

zrootlocus.m

```
% zrootlocus
% Demonstration of the root locus for discrete-time
```

```
% systems.
clf
zros=[0.3;    -0.3];
poles=[0.1;    0.5+0.5*i;    0.5-0.5*i];
gain=1;
[numd,dend]=zp2tf(zros,poles,gain);
rlocus(numd,dend)
axis('square')    % So the unit circle is round.
zgrid
```

The result of running the file is shown in Figure 8.8. As you can see from this root locus plot, two poles leave the unit circle at sufficiently high gain, k. Thus, this system will become unstable at high gain. Compare the **zgrid** lines in Figure 8.8 with the plots in Figure 8.6.

We conclude this section with file *ddampex.m*, which demonstrates how to use **ddamp** to extract damping and natural frequency information for a discrete-time system. The command **ddamp** for discrete-time systems is analogous to **damp** for continuous-time systems. First, we create a state-space realization of our four-pole system from *dimpulseex.m*. We then create a discrete-time state-space approximation to this system using **c2dm** with six input arguments. MATLAB interprets these input arguments as the A, b, c, and d matrices of a continuous-time system in state-space form, the sampling interval **Ts**, and the method to be used, **'zoh'**. We demonstrate the use of **ddamp** with two input arguments, the discrete-time A matrix **Ad** and the sampling interval, **Ts**.

ddampex.m
```
% ddampex
% Finding the damping ratio and natural frequency
% of discrete-time systems.
zros=[-.1+2*i    -.1-2*i];
poles=[-.1+i    -.1-i    -.1+5*i    -.1-5*i];
K=1;
Ts=0.3;
[Ac,bc,cc,dc]=zp2ss(zros',poles',K);
[Ad,bd,cd,dd]=c2dm(Ac,bc,cc,dc,Ts,'zoh');
damp(Ac)
ddamp(Ad,Ts)
```

The result of running the file is

Eigenvalue	Damping	Freq. (rad/sec)
-0.1000+1.0000i	0.0995	1.0050
-0.1000-1.0000i	0.0995	1.0050
-0.1000+5.0000i	0.0200	5.0010
-0.1000-5.0000i	0.0200	5.0010

Eigenvalue	Magnitude	Equiv. Damping	Equiv. Freq. (rad/sec)
0.9271+0.2868i	0.9704	0.0995	1.0050
0.9271-0.2868i	0.9704	0.0995	1.0050
0.0686+0.9680i	0.9704	0.0200	5.0010
0.0686-0.9680i	0.9704	0.0200	5.0010

The columns labeled **Equiv. Damping** and **Equiv. Freq. (rad/s)** are obtained by inverting Eqs. (8.6) and (8.7) to find the values of ζ and ω_n that would correspond in continuous time to the discrete-time eigenvalues. The column labeled **Magnitude** is the magnitude of each eigenvalue. This allows you to determine stability. If all the numbers in the column are less than one, that is, all the eigenvalues are within the unit circle, then the system is stable.

We have not demonstrated the use of either **dinitial** or **dlsim**. You should not have trouble learning to use them on your own because of their similarity to **initial**, **lsim**, and the commands we have illustrated.

8.3 Frequency-Domain Analysis of Discrete-Time Systems

The important MATLAB commands for frequency-domain analysis of discrete-time systems are **c2dm**, **dbode**, **dnyquist**, and **dnichols**. The latter three commands produce Bode, Nyquist, and Nichols plots, respectively, for discrete-time systems.

A natural way to demonstrate that these frequency-domain tools—Bode, Nyquist, and Nichols plots—are available for discrete-time systems is to begin by sampling a sinusoidal input signal. Let

$$u(t) = \cos \omega t$$

Then sampling $u(t)$ at sampling interval T_s and treating the sampled sig-

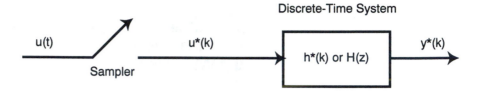

Figure 8.9   Continuous-time signal *u(t)* is sampled and the discrete signal *u* (k)* is input to a discrete-time system described by either its pulse response *h* (k)* or its transfer function *H(z)*.

nal as a discrete-time signal produces

$$u^*(k) = \cos \omega T_s k = \frac{1}{2}e^{j\omega T_s k} + \frac{1}{2}e^{-j\omega T_s k}$$

The next step is to input $u^*(k)$ into an asymptotically stable discrete-time system and to compute or measure the corresponding output $y^*(k)$. The setup is illustrated in Figure 8.9.

The computation of $y^*(k)$, based on the discrete-time convolution of the input $u^*(k)$ with the discrete-time pulse response $h^*(k)$, is

$$y^*(k) = \sum_{\ell=0}^{\infty} h^*(\ell)u^*(k-\ell)$$

$$= \frac{1}{2}\sum_{\ell=0}^{\infty} h^*(\ell)e^{j\omega T_s k}e^{-j\omega T_s \ell} + \frac{1}{2}\sum_{\ell=0}^{\infty} h^*(\ell)e^{-j\omega T_s k}e^{j\omega T_s \ell}$$

$$= \frac{1}{2}\left(\sum_{\ell=0}^{\infty} h^*(\ell)e^{-j\omega T_s \ell}\right)e^{j\omega T_s k} + \frac{1}{2}\left(\sum_{\ell=0}^{\infty} h^*(\ell)e^{j\omega T_s \ell}\right)e^{-j\omega T_s k}$$

$$= \frac{1}{2}H(e^{j\omega T_s})e^{j\omega T_s k} + \frac{1}{2}H(e^{-j\omega T_s})e^{-j\omega T_s k}$$

where we have made use of the fact that the terms in parentheses are just $H(z)$ evaluated at $z = e^{\pm j\omega T_s}$. But

$$H(e^{j\omega T_s}) = |H(e^{j\omega T_s})|e^{j\angle H(e^{j\omega T_s})}$$

and

$$H(e^{-j\omega T_s}) = |H(e^{j\omega T_s})|e^{-j\angle H(e^{j\omega T_s})}$$

Finally,

$$y^*(k) = |H(e^{j\omega T_s})|\cos(\omega T_s k + \angle H(e^{j\omega T_s})) \tag{8.8}$$

Equation (8.8) is the key to direct measurement of the frequency response of discrete-time systems as well as to defining Bode, Nyquist, and Nichols plots for such systems. For example, the Bode plots are plots of 20 log $|H(e^{j\omega T_s})|$ vs. log ω and $\angle H(e^{j\omega T_s})$ vs. log ω. To measure $|H(e^{j\omega T_s})|$ and $\angle H(e^{j\omega T_s})$ in the laboratory, we would sample cos ωt at sampling interval T_s, input the sampled signal to our discrete-time system, and measure the magnitude and phase angle of the response. We would do this for different values of ω so as to generate enough data points for accurate plots.

There is one very striking difference between the frequency-domain plots for continuous- and discrete-time systems. The frequency response of a discrete-time system is always periodic in ω with period $2\pi/T_s$. To see this you need only recognize that $e^{j\omega T_s} = e^{j(\omega T_s + 2\pi)} = e^{jT_s(\omega + 2\pi/T_s)}$. MATLAB takes advantage of this periodicity by computing and displaying the frequency response of discrete-time systems only for frequencies between $\omega = 0$ and $\omega = \pi/T_s$. You might expect the interval to be from 0 to $2\pi/T_s$. However, any interval of frequencies that is of length $2\pi/T_s$ will work equally well. In particular, the interval $-\pi/T_s$ to π/T_s will display the complete frequency response. Moreover, there is no need to display the portion of the frequency response between $-\pi/T_s$ and 0 because the magnitude is the same as that for 0 to π/T_s, and the phase angle is the negative of the phase angle between 0 and π/T_s.

We illustrate a frequency response plot of a discrete-time system with file *periodicfreq.m*. The first two lines of the file create a simple third-order continuous-time system

$$G(s) = \frac{10(s^2 + 0.2s + 2)}{(s^2 + 0.5s + 1)(s + 10)}$$

We then create two plots, shown in Figure 8.10 using the **c2dm** command.

periodicfreq.m
```
% periodicfreq
% Demonstration of the discrete-time Bode plot.
clf
num=10*[1 0.2 2];
den=conv([1 0.5 1], [1 10]);
subplot(121)
Ts=.3;
c2dm(num,den,Ts)
subplot(122)
Ts=.03;
c2dm(num,den,Ts)
```

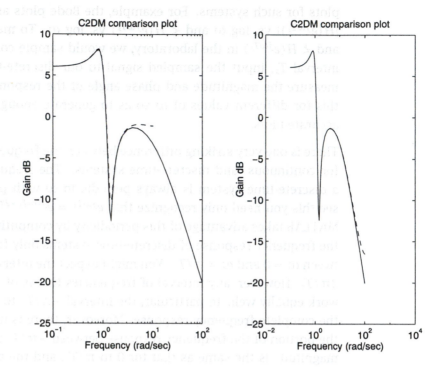

Figure 8.10   Left: Bode magnitude plots of $G(s) = 10(s^2 + 0.2s + 2)/(s^2 + 0.5s + 1)(s + 10)$ (solid line) and of the corresponding $G(z)$ using $T_s = 0.3$ second and zoh for the conversion. Right: Same as left plot except $T_s = .03$ second.

The left plot in Figure 8.10 shows a comparison of the continuous-time and discrete-time frequency responses for the example system where the discrete-time system is computed for a sampling time $T_s = 0.3$ second by the zoh method. Notice that the discrete-time Bode plot (the dashed plot) ends at $\omega = \pi/T_s = 10.5$ rad/s. It is understood that the actual discrete-time Bode plot is periodic with period $2\pi/T_s$, so you can extend the Bode plot to all frequencies from knowledge of the Bode plot between 0 and π/T_s. Because of the periodicity, you can see that $T_s = 0.3$ second is an adequate sampling interval if you are only concerned about the frequency response below approximately 3 rad/s. That is, the discrete-time Bode plot is very different from the continuous-time Bode plot at frequencies above 3 rad/s. The right-hand plot in Figure 8.10, produced by the last three lines of *periodicfreq.m*, is the same except that the sampling interval used for discretization is $T_s = 0.03$ second. Decreasing T_s by a factor of 10 extends the range of frequencies in the discrete-time Bode plot by a factor of 10, to $\omega = 105$ rad/s. Additionally, the discrete-time Bode plot with $T_s = 0.03$ second is very close to the continuous-time Bode plot for frequencies below about 30 rad/s.

To further demonstrate the periodicity in ω of sampled signals, we have created file *periodicex.m*. The file creates and plots two sampled signals on the same plot as a function of time **t**. The first vector, **y1**, contains the continuous-time signal $\cos t$, sampled every 0.3 second, for 2π seconds. The second vector, **y2**, contains the continuous-time signal $\cos(1 + 2\pi/0.3)t$, sampled every 0.3 second, for 2π seconds.

periodicex.m

```
% periodicex
% Demonstration of the periodicity of the frequency
% response of discrete-time systems.
clf
omega=1;
Ts=.3;
Tf=2*pi/omega;
t=[0:Ts:Tf];
y1=cos(omega*t);
y2=cos((omega+2*pi/Ts)*t);
plot(t,y1,t,y2)
xlabel('Time')
ylabel('Amplitude')
```

The result is shown in Figure 8.11. There are actually two plots in the figure, but they are indistinguishable because of the periodicity of sampled signals.

Our next example, provided in the file *bodesampling.m*, illustrates a Bode plot of a discrete-time system using **dbode**. We again begin by creating a continuous-time system. It is the fourth-order system from Section 8.2. We then create a discrete-time equivalent using the foh method and a sampling interval of T_s=1 second. After clearing the graphics window by means of **clf** we create a Bode plot for the continuous-time system using **bode**. The plot is shown in Figure 8.12. You should not be surprised to see the two peaks in the magnitude (at ω=1 rad/s and ω = 5 rad/s) and the two regions of sharply increasing phase angle.

bodesampling.m

```
% bodesampling
% Computing the Bode plot of a discrete-time system
% obtained by discretizing a continuous-time system.
zros=[-.1+2*i   -.1-2*i];
poles=[-.1+i   -.1-i   -.1+5*i   -.1-5*i];
```

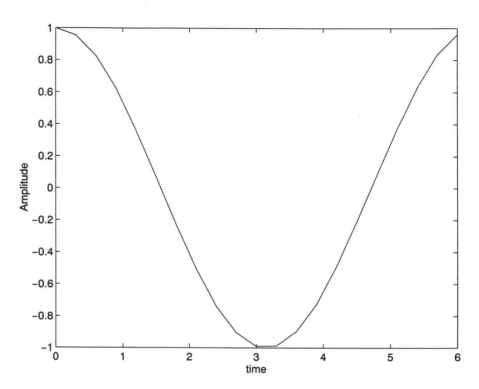

Figure 8.11 Superimposed plots of the sample values of cos*(t)* and cos$((1 + 2\pi/0.3)t)$. The sample interval is T_s = 0.3 second.

```
K=1;
Ts=1;
[num,den]=zp2tf(zros',poles',K);
[numd,dend]=c2dm(num,den,Ts,'foh');
clf
bode(num,den)
keyboard
dbode(numd,dend,Ts)
keyboard
Ts=0.2;
[numd,dend]=c2dm(num,den,Ts,'foh');
dbode(numd,dend,Ts)
```

We used **keyboard** to allow us to print the plot. We then use **dbode** to display the Bode plot of the discrete-time system created by sampling at **Ts**=1 second. Note that when using **dbode** we specify the sampling interval **Ts** as an input argument. The result is shown in Figure 8.13. Note both the greater phase shift compared to that of the continuous-

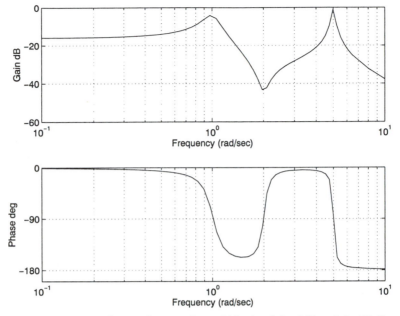

Figure 8.12 Bode plot of $G(s) = (s + 0.1 + 2i)(s + 0.1 - 2i)/(s + 0.1 + i)$ $(s + 0.1 - i)(s + 0.1 + 5i)(s + 0.1 - 5i)$.

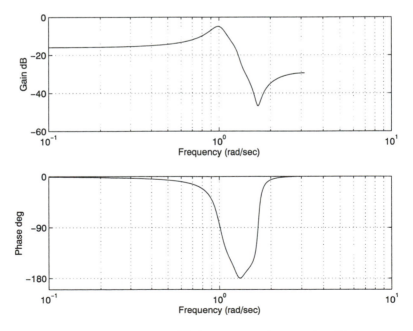

Figure 8.13 Bode plot of the $G(z)$ corresponding to the $G(s)$ in Figure 8.12 when the sampling interval $T_s = 1$ second and foh is used for the conversion of the continuous-time system to its discrete-time equivalent.

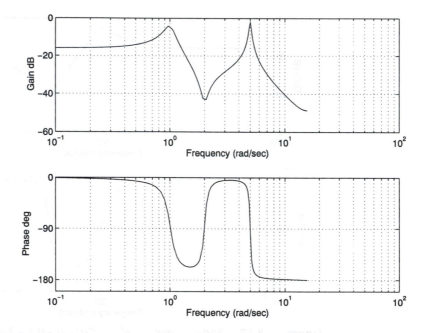

Figure 8.14   Same as Figure 8.13 except T_s = 0.2 second.

time system and the frequency at which the discrete-time signal repeats ($\omega = \pi$). We then repeat the computations for **Ts**=0.2 second to produce Figure 8.14. Note the improvement in the match between the discrete- and continuous-time phase angles for frequencies between 1 and 2 rad/s that results from faster sampling.

Our next example returns to the third-order system of the previous section and generates Nyquist plots for the discrete-time system using **dnyquist**. The file is called *dnyqex.m*. We first plot a Nyquist plot of the continuous-time system and display it in the upper left corner of Figure 8.15. We then use Tustin's method to create three different discrete-time system approximations with sampling intervals of 3, 0.3, and 0.03 second. We have discussed sampling this system in Section 8.2 in connection with file *samplingex.m*. You should review that discussion to see that T_s = 3 seconds is much too large for accurate reproduction of the continuous-time response.

dnyqex.m

```
% dnyqex
% Nyquist plots for a system sampled at several
% different rates.
```

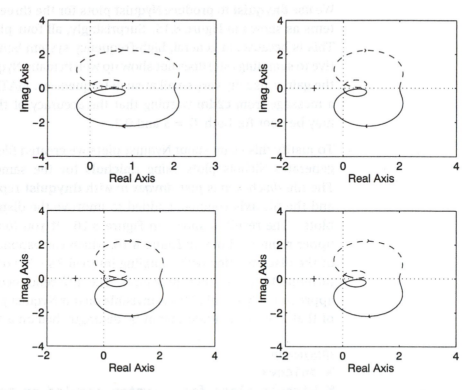

Figure 8.15 Upper left: Nyquist plot for $G(s) = 10(s^2 + 0.2s + 2)/(s^2 + 0.5 s + 1)(s + 10)$. Upper right: Nyquist plot for the corresponding $G(z)$ when the sampling interval is $T_s = 3$ seconds and Tustin's method is used for the conversion. Lower left: Same as upper right plot except $T_s = 0.3$ second. Lower right: Same as upper right plot except $T_s = 0.03$ second.

```
num=10*[1 0.2 2];
den=conv([1 0.5 1],[1 10]);
clf
subplot(221)
nyquist(num,den)
m=1;
while m<=3,
      Ts=3/10^(m-1);
      [numd,dend]=c2dm(num,den,Ts,'tustin');
      subplot(221+m)
      dnyquist(numd,dend,Ts)
      m=m+1;
end
```

We use **dnyquist** to produce Nyquist plots for the three discrete-time systems as shown in Figure 8.15. Surprisingly, all four plots look identical. This is because, in general, high-frequency system behavior (most sensitive to sampling rate) does not show up well in many Nyquist plots because the gain is usually very small at such frequencies. MATLAB does give you a message from **c2dm** warning that the accuracy of the d2c conversion may be poor for both $T_s = 3$ and 0.3.

To justify this claim about Nyquist plots we created file *dnichex.m*, which generates Nichols plots using **dnichols** for the same cases as above. The file *dnichex.m* is just *dnyqex.m* with **dnyquist** replaced by **dnichols** and the file **axis** command added to improve the display of the Nichols plots. The result is shown in Figure 8.16. If you look carefully at the upper right-hand plot in Figure 8.16, which corresponds to the response of the discretization with sampling interval **Ts** = 3, you will see that it is different from the other three plots. The difference occurs at gains below approximately –15 dB. This is invisible on our Nyquist plots because gains of that low a magnitude cannot be distinguished on a Nyquist plot.

dnichex.m
```
% dnichex
% Nichols plots for a system sampled at several
% different rates.
num=10*[1 0.2 2];
den=conv([1 0.5 1],[1 10]);
clf
subplot(221)
nichols(num,den)
axis([-360 0 -45 8])
m=1;
while m,=3,
      Ts=3/10^(m-1);
      [numd,dend]=c2dm(num,den,Ts,'tustin');
      subplot(221+m)
      dnichols(numd,dend,Ts)
      axis([-360 0 -45 8])
      m=m+1;
end
```

We encourage you to explore MATLAB's commands for studying discrete-time systems further on your own.

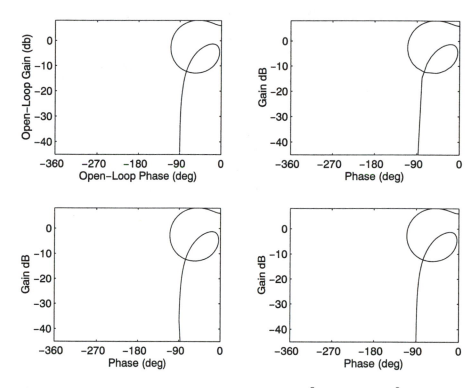

Figure 8.16 Upper left: Nichols plot for $G(s) = 10(s^2 + 0.2s + 2)/(s^2 + 0.5s + 1)$ $(s + 10)$. Upper right: Nichols plot for the corresponding $G(z)$ when the sampling interval is $T_s = 3$ seconds and Tustin's method is used for the conversion. Lower left: Same as upper right plot except $T_s = 0.3$ second. Lower right: Same as upper right plot except $T_s = 0.03$ second.

8.4 Exercises

1. (a) Repeat *samplingex.m* after replacing **impulse** by **step** and **dimpulse** by **dstep**. Compare the relations among the step responses with those among the impulse responses. Explain why the step responses match better.

   (b) Compare the response of the continuous-time system to an input $u(t) = 1 - cos(t/2)$, $t \geq 0$ with the responses of the sampled data systems $(T_s = 1, 0.1, 0.01)$ to an input $u(k) = 1 - cos(kT_s/2)$, $k \geq 0$. Discuss the similarity, or lack of similarity, of the responses.

2. (a) Compute the poles and zeros of the discrete-time systems that result from discretizing the continuous-time system with transfer function $G(s) = 10(s^2 + 0.2s + 2)/(s^2 + 0.5s + 1)(s + 10)$ for sampling intervals ranging from $T_s = 1$ second to $T_s = 0.01$ second.

(b) Create a root locus plot showing the evolution of the poles and zeros computed in part (a) as a function of T_s.

(c) Repeat parts (a) and (b) for $G(s) = (s + 0.1 + 2i)(s + 0.1 - 2i)/(s + 0.1 + i)(s + 0.1 - i)(s + 0.1 + 5i)(s + 0.1 - 5i)$.

(d) Repeat parts (a), (b), and (c) for different discretization methods.

3. (a) Transform $G(z) = (z + 0.3)(z - 0.3)/(z - 0.1)(z - 0.5 + 0.5i)(z - 0.5 - 0.5i)$ into a continuous-time equivalent system by means of the zoh method for sampling intervals ranging from $T_s = 0.01$ second to $T_s = 1$ second.

(b) Determine the poles and zeros of the resulting continuous-time systems.

(c) Plot root locus plots for each of the continuous-time systems.

Index

In this index, script file names appear in *italic* type and MATLAB commands appear in **bold** type.